CAMOUFLAGE AND ART

DESIGN FOR DECEPTION IN WORLD WAR 2

CAMOUFLAGE AND ART

DESIGN FOR DECEPTION IN WORLD WAR 2

HENRIETTA GOODDEN

UNICORN PRESS
LONDON

Unicorn Press
76 Great Suffolk Street
London
SE1 0BL

www.unicornpress.org
email: unicornpress@btinternet.com

First published by Unicorn Press 2007

ISBN 978 0 906290 87 2

Designed by Helen Swansbourne
Printed in Slovenia for Compass Press Limited

Illustrations
Cover: Eric Ravilious, *Leaving Scapa Flow*, 1940. Cartwright Hall
    Art Gallery, Bradford
Endpapers: William Murray Dixon, drawing of a dummy tank
    draped in camouflage netting
Frontispiece: Colin Moss, *Power Station*, purchased in 1943.
    Imperial War Museum
Right: Victorine Foot, *Camouflaging a Cruiser in Dock*, purchased
    in 1943. Imperial War Museum

Contents

T.E. La Dell, The Camouflage Workshop, Leamington Spa, *1940 (detail)*.

Foreword

DURING THE SECOND WORLD WAR, while the students of the Royal College of Art were evacuated to Ambleside in the Lake District – men in the Queens Hotel and women in the Salutation Hotel – many of the future Professors of the College were busy researching the design of camouflage for the Camouflage Directorate.

The students in and around Ambleside – whether they happened to be studying art or design – did not for obvious reasons have access to much heavy equipment, so they had a wonderful time becoming landscape artists instead. Meanwhile, in Leamington and Farnham, as this fascinating book points out, there were two of the most concentrated communities of artists and designers to be found anywhere in Britain, and all working on camouflage.

The Secretary to the Directorate of Camouflage at that time was Robin Darwin; and when he became Principal of the RCA in 1948, he re-assembled some of his key camoufleurs: nearly all his Design Professors (Graphics and Illustration, Printmaking and Engraving, Industrial Design, Furniture, and Jewellery) had the Camouflage Directorate in common, as did several of his tutors.

This was the generation which transformed the RCA into the recognisably modern College, which masterminded the Festival of Britain in 1951 and put British art and design on the international map in the austere postwar years. Robin Darwin's approach to making senior appointments may not have been strong on equal opportunities – he much preferred the Napoleonic system – but it was certainly effective. When asked what his philosophy of art education was, Darwin wrote in 1964.

" . . . all you had to do was to appoint the best senior staff in the country and then leave them to get on with their job without interference. I said nothing, of course, about those wayward currents of experience which throw people together . . ."

This book is about the most important of those "wayward currents of experience", the work of the camoufleurs during the Second World War. It is at the same time an informal history of camouflage in the war years, an account of what happens when artists, designers and scientists work together on a common cause, the story of a group of individuals who were as talented as they were eccentric, and a hidden history of British art. It could have perhaps been subtitled *Art and Illusion*, if Ernst Gombrich hadn't got there first. And best of all, *Camouflage and Art* has been written by the daughter of the expert on colour in the Naval Camouflage Unit – the architect and designer Robert Goodden. It has taken a member of the family to remove the layers of camouflage which have disguised all this from public view for well over half a century.

SIR CHRISTOPHER FRAYLING
Rector, Royal College of Art

Author's Preface

SOME TIME AGO, a friend and colleague suggested that she and I should record the reminiscences of our elderly fathers, both of whom had worked in Camouflage during the Second World War. Until then, I had been unaware that when the Royal College of Art reopened after the war, the Rector Robin Darwin's selection of his first team of professors and tutors had been largely based on contacts he had made in his wartime role as Secretary to the Directorate of Camouflage.

Alan Fletcher, in *The Art of Looking Sideways*, remembered the student joke at the College in the 1950s was that 'the teaching wasn't up to much as most of the staff had been in the army camouflage unit' – in his opinion, a particularly appropriate qualification. *Camouflage and Art* was inspired by the creativity of this wonderful group of designers and artists (in various branches of the Services) and my interest in the way they adapted their skills to such vital use. Far from being a thorough investigation of World War 2 camouflage, it records a mere selection of the many ingenious methods used to solve problems of disguise and deception. The parameters of my research were set by the personal recollections and records referring to a limited and diverse group of people, all practising designers and artists before the war, many of whom had strong associations with the RCA. These people worked in specific areas of camouflage, and their work defined the boundaries of my subject matter: for example, the camouflage of airfields is investigated, but not that of aeroplanes; the war in North Africa is the sole overseas operation recorded in detail here, other than the preparations for D-Day. First-hand accounts and individual reminiscences have dictated the direction of my study, which has been a source of learning, fascination and wonder.

This book is dedicated to the memory of Robert Goodden and all the Royal College of Art camoufleurs.

*Opposite: Robin Darwin,
Camouflaging the New
Flight Shed, 1941 (detail).*

CHAPTER ONE

Introduction

'Camouflage as it is practised today is at once an art, a craft and a science.'[1]

THE MEANING OF THE French verb *camoufler* is, depending on the source consulted, 'to blind or veil', 'to conceal, cover up, disguise' or more specifically, 'to put on stage make-up' or 'to play a practical joke'. In early twentieth-century Parisian slang the word *camouflet*

(literally translated from ancient French 'hot face') came to imply 'a puff of smoke', which would blind and confuse a robber's victim. The *Oxford English Dictionary* states that the first appearance of the word in Britain was in the *London Daily News* of 25 May 1917, during fighting in the First World War: 'the act of hiding anything from your enemy is termed "camouflage"'.[2]

The birth of military camouflage as we understand it took place in the devastating battles of the 1914–18 war. In 1915 the first 'section de camouflage' was established in France as the result of an idea proposed by Lucien Guirand de Scevola, an artist who was in the French infantry. He became director of all French camouflage operations, and by 1917 was in charge of a workforce of around 3,000 people (mainly artists and designers). The inspiration spread and the Battalion of Royal Engineers set up the British Camouflage Service in 1916, following a suggestion made by the portrait artist Solomon J. Solomon. He had approached the War Office with a scheme to hide trenches with screens made of butter muslin, which was tested with encouraging results. The British camouflage division operated mainly on the battlefields of France, under the direc-

Opposite: A British First World War camouflage factory, probably in northern France. These women are garnishing camouflage netting with hessian strips.
Imperial War Museum

Left: Lt Guirand de Scevola (right) with Mme Pierat of the Comédie Française and a British officer on a site visit to the Section de Camouflage, France, 1915.
Imperial War Museum

tion of Lieutenant-Colonel Francis Wyatt, and grew to a substantial 400-strong outfit with 60 officers.

Since the beginning of the First World War, British military vehicles and guns had been painted with variegated coloured patterns in order to disguise them. These patterns were derived from the study of animal camouflage, and the influence of nature on the development of all First and Second World War camouflage cannot be ignored. The theories of the American naturalist and artist Abbott H. Thayer (recorded in *Concealing Coloration in the Animal Kingdom*, published in 1909) were used as a foundation for US military camouflage in the 1914–18 war. His observations were based on the ability of animals and birds to blend in with their surroundings by various natural means: countershading, where the lower part of the body is of a lighter tone than the upper surfaces and the contrast between the two is blurred; mimicry, imitating the creature's surrounding habitat; and disruptive colouration (later sometimes to be interpreted as 'dazzle' camouflage), breaking up surface continuity and confusing the image perceived by the viewer. Thayer was brought in as a consultant to both the American and the British Forces. He advised the British Admiralty to paint its battleships in tones of white and pale grey, a colour theory formed by learning from the passenger liner *Titanic*'s disastrous collision with an iceberg which had been invisible to the ship's lookout. He also suggested that the British army field dress, a plain khaki originally adopted to blend into a background of dusty Indian terrain, be changed to a disruptively patterned uniform.[3]

During this period the urgent necessity of hiding all equipment and ammunition became a priority, and camouflage systems were developed in order to achieve this. Although the British military authorities were reluctant to employ artists (an ironic demonstration of the difference in French and British attitudes!), their skills together with those of scientists in the Allied troops were used to invent new ways to deceive and confuse the enemy. At first, the two most common methods of deception were gun-concealing nets made from fishing net 'garnished' with bunches of dyed raffia (later to be replaced by painted canvas strips), and huge canvas sheets daubed with random splodges to blend in with a natural landscape, which were mainly created to cover ammunition dumps.

Early in 1916, after a year during which light-seeking Zeppelins posed a constant threat, it was suggested by the Royal Naval Air Service (in control of Britain's air defences) that dummy airfields 'laid out with flares', and illuminated decoy towns, would be an effective way of baiting airships to an area where they could then be targeted by fighter planes.[4] Although there were two occasions when this plan was successful it did not become common practice, mainly because the Admiralty's responsibilities for domestic air defence were transferred to the Royal Flying Corps which appears not to have used dummy airfields. On the Western Front in 1918, however, false flarepaths and lit-up dummy buildings did play a successful part in deflecting German night attacks from their real targets. The French also created dummy decoy airfields for

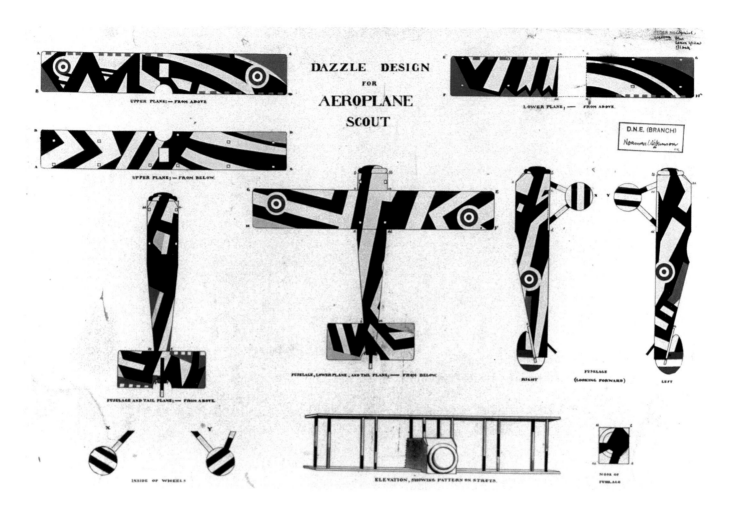

Norman Wilkinson's dazzle design for an aeroplane. His dazzle theory for ships is well-documented, but it is unclear how this would have worked on aeroplanes. Ink and watercolour on paper. RAF Museum

daylight use, their success being a consequence of the ease in emulating the very basic and simply designed character of the genuine installation.[5]

The 'dazzle' style, so-called after a pre-war American expression 'razzle-dazzle' (in French *zebrage*), featured prominently during the First World War, and is probably the most universally-known form of naval camouflage. It was memorably documented by the Vorticist painter Edward Wadsworth, whose wartime remit was to supervise the patterning of around 2,000 ships.[6] From March to December 1917, German submarines succeeded in sinking a shocking total of 925 British ships. Naval Lieutenant Norman Wilkinson (in peacetime a designer and marine painter) introduced the idea that abstract patterns of colour could cause visual confusion and therefore affect the path of

14

torpedoes fired from a submarine by means of observation through a periscope. A dazzle department was set up under Wilkinson in the basement of the Royal Academy of Arts, where model boats were painted (by 'female art students') with disruptive designs in highly-contrasting tones of pale colours as well as blacks and greys.[7] Participating artists included Wadsworth, C. H. R. Nevinson, David Bomberg and William Roberts. The angular shapes and blocks of colour were intended to distort a ship's outline and to allow the vessel to blend in with its surrounding seascape. These schemes were carried out in the dockyards of Bristol, Cardiff, Liverpool and Southampton. However, their results were viewed with caution. A document written in 1942 by the Director of Staff and Training Duties (DSTD) states that 'camouflage for confusion of inclination or identity (dazzle painting) is not considered to have been a great success in the last war'. As an interesting contrast, the Admiralty's report on the Committee on Dazzle Painting published in 1918 suggested that the practice could be continued as it was 'not disadvantageous', and had resulted in an increase in the confidence and morale of crews on dazzle ships.[8]

When the War Office set up its Camouflage Development and Training Centre at Farnham Castle in 1940, the eminent naturalist Dr Hugh Cott was recruited as an essential part of the team of instructors. His book *Adaptive Colouration in Animals* was a compulsory text for those involved in camouflage, and his work continued that of Abbott H. Thayer. A review in the British journal *Nature* (1941) by the biologist John Graham Kerr, refers to Dr Cott's book as 'the greatest comprehensive work on the subject [of natural camouflage]'. The principles of countershading, mimicry and disruptive colouration were to be adopted again and again in Second World War camouflage schemes.

Kerr was a well-known biologist and Regius Professor of Zoology at Glasgow University. He was one of a substantial group of Establishment figures who purported not to understand the recruitment of trained artists in camouflage. In a protracted speech in the House of Commons in 1941 he appealed to Members to listen to his views as an expert in natural camouflage, and to agree with him that the War Cabinet (which had been ignoring his repeated private representations) should use biologists, not 'physicists or artists', to head up all camouflage research and design work. He recalled work in the First World War where he had been called upon as a consultant, and gave lengthy and detailed examples of natural camouflage and its applications. His grievance was that 'those responsible for [current] camouflage' were either totally ignorant of these scientific principles, or that they defied them completely.[9]

'Nature invented the art of camouflage, and man has developed it as a science of modern war.' Lieutenant-Colonel C. H. R. Chesney's book *The Art of Camouflage* (1941) is introduced by another biologist, J. Huddlestone, whose view on the subject is that behind the execution of a successful camouflage scheme are the 'scientist and the research chemist, skilled in the actinic properties of colour, and the suitability of all kinds of materials'. His next sentence disapprovingly states that 'Camouflage is definitely not a sphere to be invaded by amateurs. What was once merely a matter of the casual daubing of guns and other objects of military importance with paint, has now become a highly specialized science.' Huddlestone also goes on to remark scathingly that 'why a successful painter or scenic artist who has perhaps never been in an aeroplane should be thought ... suitable ... to undertake work of this nature, is rather astonishing'.

Chesney himself is of the more understanding opinion that 'camouflage is not *primarily* the job of a painter-artist'. He admits that 'the expert in painting has the best sense of colour, perspective and tone', but that the artist should not be placed in charge of camouflage operations: 'the more trained and expert he is as a painter, the less desirable is it that he should be placed in the position of controller'. The artist being 'a worshipper at the shrine of the paint-brush' and also being one trained to produce two-, not three-dimensional work, makes it clear to Chesney that the ideal controller of a camouflage team would be an engineer or architect, with artists carrying out the approved schemes.[10]

Lieutenant-Colonel Chesney endorses his approval of involving architects in camouflage design by adding an unusual appendix to his book. This reproduces a paper written by an unnamed architect in February 1940 (only his initials C. G. A. are given), who is unknown to the author. It is oddly laid out. The pages have been divided down the middle, with the architect's document on the left, and Chesney's comments (he stresses that these were allowed) on the right. It would appear that generally he is in agreement with the architect's views on camouflage design and construction, albeit in rather a condescending tone, but being a military man he obviously approves of the rigorous training required to become a qualified architect.

'Camouflage is the art of deception applied to vital targets which the enemy might like to attack ... [it] is the adaptation of sound scientific theory to meet particular demands and its application with judgement and commonsense by those specifically trained to use their eyes.' In February 1943 Robin Darwin, former professor of fine art at Newcastle University, and now secretary to the Camouflage Committee, wrote a letter to the Public Relations department of the Ministry of Information. He enclosed some personally written 'notes on the role of Artists in Camouflage', in answer to a request from the Ministry. These 'notes' amount to an erudite and eloquent paper explaining in the most perceptive terms why it was important for designers and artists to work alongside the 'engineers, architects and scientists' involved in the creation of successful camouflage. Darwin, being a trained painter himself, and very well-connected to the Establishment (he was descended from Charles Darwin the naturalist) was able to see the issue from both points of view, and could understand the reluctance of certain parties to accept or even try to understand the involvement of artists and designers in camouflage.

Darwin's paper begins by stating that all potential targets must be viewed from the air before any

15

Letter from Robin Darwin to the Ministry of Home Security lending his support for artists working in camouflage.

camouflage designs are begun (thus partly endorsing J. Huddlestone's disapproving reference to painters and scenic artists who have never flown).[11] He goes on to explain in some detail the methods used by the camouflage team in recreating in dummy form the conditions of the target, before emphasising the reasons why the designer needs an artist's eyes to help devise the 'best and most economical ways of applying camouflage treatment'. The breaking-down and disruption of 'those characteristics which reveal the form of an object' demand an intimate knowledge of form, as possessed by an artist who is trained to analyse and memorise what he sees and who has 'powers of observation and visual retention far above the normal standards'.

Although Darwin diplomatically emphasises that the skills of other experts must be used to assess all the factors involved, it is the artist's responsibility to contrive 'the very best concealment possible within the limits allowed'. He understands the difficulty experienced by certain government officials in having to deal with 'a race of carefree individuals such as they have never met with before', and concludes his paper with the observation that both parties have learned to work together by understanding 'in some measure' the two points of view. His words highlight the significance of the crucial relationship between art and science which underpinned the extraordinary inventions, deceptions and decoys that materialised as camouflage during the Second World War.

One of the many acquaintances made by Darwin in his post as secretary to the Camouflage Committee was

*Ruskin Spear: Robin
Darwin as Rector of the
RCA, with visual
references to both the
Establishment and
modern art, 1961.
Oil on board.*
Royal College of Art Collection

18

*Robin Darwin: Hugh Casson in his war-issue duffel-coat, which he
continued to wear well into the 1950s, 1957. Oil on canvas.*
Royal College of Art Collection

Hugh Casson, who was to become professor of the department of interior design at the Royal College of Art in 1951. In an illustrated article entitled 'Art by Accident' in *The Architectural Review* (September 1944) he remarked: 'Camouflage is not magic,' and went on to say that contrary to popular belief 'particoloured patterns, violently contrasted, confer no mantle of invisibility upon the object to which they are haphazardly applied'. The primary aim of camouflage, in his view, was to deceive the eye of a bomber pilot 'in such a way that recognition of the target is delayed or prevented for as long a time as possible'.

Casson's list of the methods by which an experienced 'camoufleur' should disguise or deceive is a brief but perceptive summary of what was needed in the Second World War:

● Intelligent siting of new buildings and man-made environments.

● Total concealment – difficult to achieve, only appropriate for top-priority subjects or objects small enough to be dealt with in this way.

● Merging or toning down – eliminating light-reflective surfaces, applying organic 'local' pattern and tone to the target.

● Disruption – breaking up the image of the surface by 'strongly contrasted tones'.

● Imitation – dummy housing, roads and other features, created in two or three dimensions.

● Decoy – 'a sideline of camouflage' – replica installations away from the real target, false flares, fake troop movements, and many other ideas.

Cover of Concealment of New Buildings, *published as an aid to camouflage design.*
HM Stationery Office

The work of the camouflage divisions set up within the various Forces during the Second World War, and the enterprising and often amazing ways in which the methods listed already were interpreted and put into action, bear out Robin Darwin's strong belief that 'so far from being black magic', successful camouflage resulted from the clever application of art to science. The development of the camouflage units and, more importantly, the genius of their protagonists, are the subject of this book.

21

Opposite: Hugh Casson, watercolour of a grass-roofed control tower with a disruptively-treated tanker. The picture shows typical concealment methods used in the interests of camouflage. © Hugh Casson, 1943. Reproduced with permission of Sir Hugh Casson Ltd

Right: See page 103

CHAPTER TWO

Government Initiatives

'FAILURE TO APPRECIATE the peculiar characteristics of the airman's view of the world below him under all conditions, makes camouflage absolutely impossible.' Julian Trevelyan's words in his expressive summary 'The Technique of Camouflage', published alongside Hugh Casson's assessment in *The Architectural Review* in September 1944, clearly point out the importance of the Air Ministry's responsibility for arranging camouflage measures to protect the country in the ever-increasing danger of war.

In October 1936 the Committee of Imperial Defence had made a recommendation that the Home Office should take charge of every matter relating to the security of all vital British 'factories, key points and landmarks' in the event of a war.[1] As attack from the air became more probable, it became imperative to research and perfect methods of concealment and disguise which would deceive and confuse the enemy's aerial observation units, and it was a natural step that the Air Ministry should eventually be given overall responsibility for these essential protective duties.

Towards the end of 1938, after the Munich Crisis and with war a menacing possibility, it seemed clear that all major industrial plants should immediately be protected by camouflage measures. Three crucial factors relating to each potential target were highlighted by the Government in a special report: 'First: probability and probable type of attack, second: importance of target to the war effort, and third: the extent to which the target can be made less conspicuous.'[2] A new operational research section dubbed RE8 [Ministry of Home Security Research and Experiments department], and attached to the Battalion of Royal Engineers, was set up by the Ministry of Supply as the unit responsible for the camouflage of all 'Military Establishments, Fixed Permanent Defences, Royal Ordnance Factories (excluding Agency factories run by civil firms) [and] Ministry of Supply Establishments'.[3] Technical advice was provided by the Camouflage Research Establishment at Farnborough (set up in December 1937 under the direction of camouflage veteran Lieutenant-Colonel Francis Wyatt) and by the Ministry's official Paint Research Station at Teddington.

The Air Ministry had already formed a section which would apply itself solely to RAF camouflage requirements, dealing with Air Force buildings and

Opposite: Colin Moss, Camouflaged Roofs, c.1940. A factory tower has been painted with a disruptive pattern for camouflage purposes, and the 'sawtooth' roof disguised with netting. Glass windows had to be treated to dull the tell-tale glare. Imperial War Museum

24

equipment. In addition to this, a further decision was made to create a section to handle civil camouflage. Adastral House in Holborn, the RAF's London headquarters, was selected as the base for the new camouflage design department, with Captain L. M. Glasson in place as chief camouflage officer. Crucial installations including power stations, gasworks, oil and water reservoirs, docks and railways, as well as major munitions factories, assembly plants and other industrial sites, were listed and classified according to their need for protection.

However, by December 1938 the threat of war appeared to have decreased, and the Civil Camouflage Section used this lull to begin preparing special designs for these 'key sites'. A large room at Adastral House was provided as a studio, with an adjoining 'viewing room' for testing models in simulated conditions. At the same time the Air Ministry set up a special department whose function was to list its most vital and important contractors. This became known as the Key Points Intelligence Branch (later Directorate) and its remit depended on reports and requests from the various Ministries. Plans and photographs of crucial sites (such as the factories of Rolls Royce, Vickers-Armstrong and Vauxhall Motors) would be submitted to the chief camouflage officer and, after obtaining a report on the tactical aspect from the Office of Air Staff, it was then the responsibility of the Camouflage Section to design and plan appropriate schemes.

Staff recruited to the creative division of the Civil Camouflage Section came from a mixture of back-grounds. It was important that the core team contained Forces personnel, particularly those with experience of flying, and an engineering or science training was considered indispensable. The December 1938 edition of the publication *Art and Industry* goes further and observes that 'camouflage requirements should include the industrial designer's knowledge of lighting and temporary structures in addition to the artist's appreciation of the appearance of things'.[4] An additional feature from the November 1939 issue comments that the 'problem' of camouflage is 'perhaps more happily dealt with by the co-operation of the physicist with a particular knowledge of problems concerning light, the constructional engineer, and the artist with a special knowledge of colour as applied to large surfaces'.[5]

The Central Institute of Art and Design (founded by a distinguished panel which included Kenneth Clark, director of the National Gallery, and Jack Beddington, advertising director for Shell-Mex) worked with the Government to compile a register of artists who would be eligible for special war work. In addition to selecting official war artists, the register advertised positions in camouflage, propaganda and publicity. Amongst the first officers to be recruited to the Camouflage Section were painters Thomas Monnington (later to become president of the Royal Academy of Art), Captain G. B. Solomon, who was nephew of Solomon J. Solomon, Cosmo Clark and Richard Carline. Successful applicants for the first round of job allocations in 1939 were painter Edwin La Dell, sculptor Leon Underwood and

The Department of Camouflage photographed in September 1944. Front row, from left to right: Gilbert Solomon, Captain Lancelot Glasson, Henry Hoyland. Back row, from left to right: L. J. Watson, G. Grayston, James Yunge-Bateman, Christopher Ironside, ? Coupland, 'Johnnie' Walker.
Property of Virginia Ironside

25

designers Richard Guyatt and Christopher Ironside. Although 'no formal decision was made to employ artists in camouflage development, there was a natural partnership' based on their aptitude for good visual recall, and their understanding of scale, colour and tone.[6] The artists' register was inundated, and on 5 October 1939 *The Times* recorded that no further applications for enrolment would be considered; numbers of would-be camoufleurs were 'sufficient for present needs'.

As work continued at the Air Ministry on the development of specific camouflage schemes, it became increasingly obvious that each of the Forces should have its own camouflage section. The issues and problems were so diverse that no single organisation was equipped to handle them all. Early in 1939 the Home Office set up the Camouflage Sub-Committee which included representatives from the Admiralty, the War Office, the Air Ministry and the Office of Works; its target was to approve a definitive general guide to camouflage design which would work for each of the Forces, and to monitor essential issues such as the specific camouflage paint colour range which was kept under review by Lieutenant-Colonel Wyatt's research department.

With war imminent, the Civil Defence Act (passed in July 1939) allowed the Camouflage Sub-Committee to insist on essential camouflage work. A Camouflage Advisory Panel which included, as well as senior members of the Forces, 'Mr Paul Nash (artist), Mr Jowett (principal, Royal College of Art)' and 'Dr Cott (biologist, Cambridge)', was appointed on 2 August.[7] After a shaky start (three senior members resigned on grounds that their research and advice was not being taken seriously), official reports confirmed that the Panel had made 'satisfactory progress under great difficulties'. On 28 August an emergency instruction for camouflage was issued to all key points, and on 2 September when war was finally officially declared, urgent telegrams were sent out calling for compulsory action on these instructions.

At this crucial moment, the Ministry of Home Security which was responsible for all areas of civil defence, was formed. The Camouflage Section was taken over by the Home Office, and in March 1940 a Camouflage Policy and Organisation Committee was established with senior staff and representatives from each of the Forces. Shortly after this its headquarters were evacuated to Leamington Spa. The newly-christened Civil Defence Camouflage Establishment set up its offices in the Regent Hotel in this elegant Georgian town, with the creative department situated at the disused skating-rink. Nearby Baginton airport was the base for aerial surveys. Dr W. E. Curtis, CDCE's scientific advisor, was appointed chief superintendent, while Captain Glasson was relocated to Leamington as chief camouflage officer. Policy, finance, supply and administrative sections remained in London.

In the summer of 1940 the Government suggested that for reasons of economy, all camouflage sections should be reunited into one department, which would also undertake the training of officers belonging to the various Services. This idea was unanimously rejected on the grounds that one establishment could not possibly solve the huge variety of camouflage issues and problems. However, a single Directorate was formed to combine the CDCE with the administrative and finance sections, all to be located at Leamington under the direction of Wing Commander T. R. Cave-Brown-Cave. The administrative division was responsible for securing three tenders for each key site to be camouflaged, and for arranging labour and technical supply. The finance section would approve and manage the consequent expenditure.

Captain Glasson then proceeded to recruit 'those having good artistic training and in many cases some experience of industrial design'.[8] By December 1940 staff numbers at Leamington amounted to 'some 84 technical officers with a subordinate staff of 20 and a clerical staff of 23', these numbers increasing as camouflage activities became more and more vital to the war effort. The gracious Regency spa was suddenly

Anne Newland, The Rink at Leamington, *ink and wash sketch, c.1942. These female junior technical assistants are working on designs and model buildings. Newland herself was a 'JTA'.*
RAF Museum

peopled with an unfamiliar group of strange individuals whose informal uniform was 'the dark blue donkey jacket'.[9] An excerpt from *The Fortnightly DO – a Bulletin for Unit Concealment Officers* quotes (somewhat facetiously to begin with) that 'an old skating rink in Leamington is now filled with artists with hair of various lengths, painting camouflage schemes on to models of all the most important factories engaged in

our war production … it is not only artists, but scientists, engineers and photographic specialists … that are contributing to this work'.[10]

In 1941 a new Camouflage Committee was set up with Cave-Brown-Cave as chairman, and former Leamington recruit Robin Darwin was appointed as secretary. A Technical Sub-Committee was also formed as essential back-up. The most valuable feature of

28

```
        DIRECTORATE OF CAMOUFLAGE - NUMBERS & GRADES

                    OF STAFF.

                    1 Director

Technical.                          Administrative & Finance.

1  Chief Camouflage Officer.        1  Principal.

1  Senior Design Officer.           1  Senior Staff Officer.

1  Deputy Senior Design Officer.    4  Staff Officers.

1  Senior Constructional Officer.   10 Executive & Higher Clerical
                                       Officers & Temporary Assistants.
1  Deputy Senior Constructural
            Officer.                16 Clerical Officers & Clerical
                                                       Assistants.
30 Camouflage Officers.
                                    48 Clerical Officers (Temporary).
1  Principal Technical Assistant.
                                    11 Shorthand Typists & Typists.
15 Senior Technical Assistants.
                                    2  Telephonists.
20 Junior Technical Assistants.
                                    7  Messengers.
10 Technical Officers.
                                    5  Night Watchmen.
12 Assistant Technical Officers.
                                    15 Cleaners.
3  Smoke Sub-Section.
                                    ───
5  Horticultural Sub-Section.       120

2  Map Section.
                                    LONDON.
4  Photographic Section.
                                    1  Principal.
1  Storekeeper.
                                    1  Staff Officer.
1  Packer.
                                    5  Clerical Officers & Clerical
4  Labourers.                                          Assistants.
                                    ───
───                                 7
113

            TOTAL   STAFF   241
                           ───

        Staff of Research & Experiments Branch under a
        Scientific Adviser (attached to Camouflage
        Directorate) number  28.
```

Leamington staff list dated 6 April 1941. Note that the scientific research staff are listed separately.

these groups was the wide discussion of camouflage techniques. With the growing understanding (based on experience of combat in North Africa and the Middle East) that camouflage should involve a knowledge of three-dimensional structure, a further decision was made to recruit architects and engineers. It seems, however, that on the formation of the Directorate the year before, the Chief Scientific Adviser, Dr Reginald Stradling, had short-sightedly chosen to remove Dr Curtis and his experts to another site (away from the bohemian ideas of the artists) in order that they should deal independently with scientific problems. This naturally made the Directorate's work much more difficult, as endorsed by Cave-Brown-Cave's comment that he 'would have been greatly assisted by one or two scientific officers attached to his Directorate'. This problem was evident from time to time in the disbelieving and sometimes insulting attitude of certain senior members of the Government and the Forces to the idea that artists and designers were useful and effective in camouflage.

The preface to the Ministry of Home Security's document written at the end of the war to record the activities of the Directorate of Camouflage mentions that 'a short course at Leamington was included in the training of Army Static Camouflage Officers', and also that 'lectures were given … at various Regional Headquarters'. The War Office pointed out that the training of officers in camouflage was a question of applying the principles to military requirements, and should be co-ordinated with military training, as was happening on 'special courses for the purpose'. There must have been some discussion about transferring the training courses to Leamington, but this resulted in a firm and pertinent statement that the headquarters did not have the staff or equipment to deal with the military aspect, and that the official training establishments' instructors should continue to maintain 'effective liaison' with the civil camouflage organisation. The same would apply to the Admiralty – it would continue to uphold the closest liaison with the civil camouflage organisation while selecting and training its camouflage officers under its own steam.

In January 1942, perhaps with hindsight resulting from the devastating air-raids of the Blitz, the real work began. The Civil Camouflage Assessment Committee began systematically to assess important potential target sites, and it was decided that 750 out of a total of 2,550 had urgent need of protection. Two hundred were inspected for provision of daytime concealment, the remainder for night-time only. A table of priorities was drawn up, and the Key Points Intelligence Directorate prepared a classified list of sites in order of supply importance. A high proportion of these key sites were large working factories, and the security and morale of staff were naturally of prime importance to essential wartime supply. Many of the enemy's camouflage arrangements were evidently designed with reassurance of factory staff as a main priority, although these schemes (according to British observers) had very little operational value. However, although no British camouflage treatment was carried

29

Date July 11 42

~~Strictly Confidential~~

HANDBOOK OF
CAMOUFLAGE PRACTICE

ISSUED BY THE
Directorate of Camouflage
MINISTRY OF HOME SECURITY
LEAMINGTON SPA

Robin Darwin's copy of the official Handbook of Camouflage Practice, July 1942. The words 'strictly confidential' were obliterated at a later date.

Stanley William Hayter working on a model for the ICRU, 1940.
Photograph by Lee Miller © *Lee Miller Archives, England, 2006. All rights reserved*

out solely for its effect on morale, workers were no doubt reassured by its existence. As the output of the Key Points Intelligence Directorate grew, so did the confidence of the workforce and the morale of the general public, who realised that expert and careful consideration was being put into action in its defence.

Early in the war, in advance of all this official activity, an unofficial group calling itself the Industrial Camouflage Research Unit had been set up as an independent camouflage consultancy. Originated by Stanley William Hayter on the strength of his camouflage experiences in the Spanish Civil War, its members included fellow artists Roland Penrose, Julian Trevelyan, Helen Phillips and John Buckland Wright, and Denis Clark-Hall, an architect.[11] The group rented a space in the Bedford Square offices of the architect Erno Goldfinger, and began by addressing the concealment of buildings and sites against ground-level attack. One of its methods of working, where possible, was to hire planes and take photographs from the air in order to learn about visual deception.[12] Trevelyan later explained in *The Architectural Review* how 'the vertical air photograph can give such an accurate map-like picture that even the minutest disturbances of the ground can be examined closely, and ... the relief, form, texture and nature of most features can be accurately determined'.[13] The team made models, very simplistic at first ('green wiggly lines'), and also experimented with various primitive types of deceptive treatment such as masking tractors with seaweed. Its work became more

Screens for a full-scale model of anti-tank gun camouflage, with ICRU members, 1940.
The image shows similarities to the model developed by S. W. Hayter.
The Mastes and Fellows of Trinity College Cambridge

32

Victorine Foot, pen-and-ink sketch of Robin Darwin, 1941. Foot was a JTA and a prolific artist.
Imperial War Museum

professional as methods of deception using industrial materials such as asbestos sheeting and strip metal were developed, and it is not surprising that as a result of the team's increasing renown it was retained by a number of private companies and individuals to develop unique camouflage schemes.

The Industrial Camouflage Research Unit kept up its experimental work throughout the 'phoney war' (the relatively quiet period between the invasion of Poland in 1939 and the fall of France in 1940), unaware that the Government had by then set up the CDCE in order to protect all vital factories. The authorities clamped down on the unofficial unit's requests for flights and as a result of this it was unable to continue its airborne observations and consequently the creation of its designs. While waiting to go into service after the fall of France in 1940, Julian Trevelyan decided to try, on the strength of his ICRU experience, for the Camouflage Section. He was interviewed and accepted by Colonel Richard McLean Buckley as a recruit for the War Office's new Camouflage Development and Training Centre (CD&TC) at Farnham Castle, Surrey.

A letter from an official at the War Office to a member of staff at the Treasury, written on Boxing Day 1940, underlines the fact that 'while it might be possible to have the basic principles taught at Leamington, the methods of applying those principles to Army needs must … continue to be taught at Farnham'. The first Camouflage Development and Training Centre had been set up at Farnham Castle in May 1940 as a hasty reaction to the War Office's realisation that the Home Forces were totally unequipped to meet an impending invasion. Its director was Colonel Frederick Beddington, brother of Jack Beddington, and the Centre absorbed much of the work of the RE8 research department,

becoming a model for all military camouflage training throughout the war.

As officers graduated from this course, they would almost always become army camouflage instructors in the Battalion of Royal Engineers, and were in turn active in establishing similar courses at strategic points – for example the Eastern Command Camouflage School in Norwich, South-Eastern Command in Tunbridge Wells, and Northern Coastal Command based near Darlington. The Home Guard had its own training school at Osterley in Middlesex, run by the artist Roland Penrose. A Scottish school was set up in Edinburgh. Many officers of the Royal Engineers spent the early part of the war in the Middle East, and their training school at Helwan in Alexandria was staffed by officers trained at Farnham. Alongside the distinguished team attached to Leamington, the list of artists, designers and architects who completed their training at Farnham and went on to create ingenious methods of deception and decoy throughout the war is astonishing. It represents nothing less than an impressive catalogue of many of the strongest and most eminent practitioners of mid-twentieth century art and design.

33

CHAPTER THREE

Training Camoufleurs

WHEN COLONEL FREDERICK BEDDINGTON was commissioned in spring 1940 to establish the Royal Engineers' Camouflage Development and Training Centre, eventually to be based at Farnham Castle, he recruited Colonel Richard McLean Buckley as chief instructor to the first training course. Buckley was ideally suited to this position, having experienced camouflage work under Solomon J. Solomon on the battlefields of France and Flanders in the First World War. Beddington's rationale was that camouflage was a visual art, and recruits were selected accordingly. Amongst the first cohort of trainees to join No. 1 Camouflage Course (based at Netheravon Barracks and later moved to the Royal Engineers' accommodation at Aldershot) were old friends from the Slade School of Fine Art, including the artists Frederick Gore (son of the painter Spencer Gore), Lynton Lamb, Blair Hughes-Stanton and Edward Seago, as well as Julian Trevelyan and the sculptor John Codner. Designers James Gardner and Ashley Havinden were brought in alongside Oliver Messel, the famous theatre designer. Magician Jasper Maskelyne was selected for his skills in deceiving the eye. The zoologist Dr Hugh Cott and

Francis Knox, who was the most senior recruit, were selected for their expertise in animal colouration. Later postings included such distinguished names as the architect Basil Spence and the typographer John Lewis.

Also present on the first camouflage training course, in the role of observer, was the surrealist artist Roland Penrose, who had been a member along with Julian Trevelyan of the Industrial Camouflage Research Unit. Penrose, who was married to the American photographer Lee Miller, was a civilian lecturer at the War Office School for Instructors to the Home Guard at Osterley Park in Middlesex, for which he wrote the *Home Guard Manual of Camouflage*.

This unusual and challenging group of recruits was required to complete a short course of military duties to prepare themselves for six weeks of training at Farnham, which in turn would equip them to become staff officers (camouflage) attached to army or corps headquarters. Many of the team were the least soldierly characters imaginable. James Gardner, for example, writes in his memoirs that when he bought his off-the-peg army uniform, he unwittingly fixed the badges 'the wrong way up'. His recollection of the preliminary

'Device 91' (an inflatable tank developed at Farnham) advancing through rural Surrey lanes, 1943.
Property of David Medd

Farnham 5173 Ext 3.

Subject:- Personal Appearance.

Ref:26/G/41/3361.

G.S.O.2 (Cam),
    G(Ops), C.H.Q. Home Forces.
H.Q. Northern, Eastern, S.Eastern, Southern,
    Western, Scottish, Anti-Aircraft Command.
H.Q. Second Army.
G.S.O.3(Cam),
    H.Q. Northern Ireland District,
    London District.
H.Q. 1st Corps, 12th Corps.
Camouflage Officer, School of Infantry.

----

1.   It has been brought to our notice that there has been
a lack of smartness and tidiness in personal appearance
noticeable amongst certain Camouflage Officers.

1.   In as much as it is accepted that the smart soldier
is the good soldier, and in as much as we must identify
ourselves with the activities of the soldier, it is important
not to prejudice our connections with senior commanders by
slackness of this description.

3.   Will you take steps, please, to ensure that all
Camouflage Officers in your Command are shown this letter
and that they act on it ?

                          R.H.KELLY.

                          Major R.A.,
                          A/Commandant,
                Camouflage Development & Training Centre, R.E.

Farnham Castle,
FARNHAM, Surrey.

22nd July, 1943.

Memo referring to the camouflage officers' lack of sartorial correctness.
Imperial War Museum

military training is of being 'shown how to salute' in order to at least look like an officer![1] A letter dispatched to all camouflage headquarters later in the war from Major R. H. Kelly at Farnham, points out that 'it has been brought to our notice that there has been a lack of smartness and tidiness in personal appearance noticeable amongst certain Camouflage Officers', and asks that the letter be shown to all relevant personnel. Major Kelly goes on to say, 'the smart soldier is the good soldier.' It may have been artistic individuality rather than scruffiness that was upsetting the commandant!

Colonel Buckley found the experience of teaching the group a considerable challenge, remembering a feeling of great relief that a team of men so unprepared for the strictures of military training had managed to complete the course without much more than the odd minor injury.

Julian Trevelyan was not required to be present on the course of military duties, owing to his previous camouflage experience. He remembers that the Farnham trainees were 'an amusing sort of crew', learning the required skills and then 'sent out to preach camouflage'.[2] In the words of Major Kelly 'the art of instruction is that of maintaining the interest of the students'.[3] The lecturers used a certain amount of play-acting to catch the imagination of their audience, who as 'arty types' were expected to respond to this kind of approach. Trevelyan's evaluation of the trained camouflage officer's main focus is that 'in all the services much … time is spent in explaining to all ranks the peculiarities of air view, in the hope that once this has been understood the rest will follow'.[4]

The remit of the Training and Development Wings is laid out in the 'Charter, Camouflage Development and Training Centre, RE'.[5] It specifies the ways in which camouflage officers should be trained and kept up to date, and also lists the duties of the Development Section. As well as courses for staff officers (camouflage), the Centre ran training courses for senior Royal Engineers officers – battalion commanders and, later, formation commanders.

Observing the inventive and effective ways they

If the enemy airman sees **THIS . . .**

you will get **THIS**

Spoil
Tracks
through
the
encircling
Wire
pointing
to
Pl. H.Q.

Remove this if possible

dealt with camouflage issues, Colonel Buckley soon made the decision that several of the group should themselves be involved in teaching. Some, like James Gardner, were kept on at the CD&TC as instructors once the training course was over, others were posted to Northern, Eastern, South-Eastern or Southern Command to work directly with regimental instructors. Many Royal Engineers went directly to the Middle East for active service with the 8th Army, to create camouflage schemes which were to have a bearing on issues of decoy and deception throughout the rest of the war.

Gardner states that he was put in charge of demonstrating a broad range of practical techniques. Many types of field camouflage were devised, and these needed to be described to troops in an approachable and easily understood manner, resulting in a series of booklets such as *Concealment in the Field*. In addition to Gardner's graphic output, the commercial artist Ashley Havinden must have been responsible for many of the instruction books and brochures issued to the staff – they portray a state-of-the-art style in graphics, presumably executed on a shoestring budget. Havinden's very comprehensive archive contains clear and wittily annotated lecture notes for an NCO's training course (No. 5, held on 12–13 November 1941). He also made copious notes at a series of refresher courses delivered by Colonel Buckley for camouflage officers in 1942, which show how lively and thought-provoking the lectures must have been. The main questions, summarised in the notes, were:

1. Against what form of observation or attack am I to hide?
2. What will the observer or attacker see?
3. Why will he see it? (ANALYSIS)
4. As what shall I conceal it?
   a) As nothing (i.e. merging)
   b) As something innocent that will excite no interest (disguise)
   c) Consider value of distracting away from post [position] by misdirection.

Havinden's notes point out that very short courses and demonstrations were not effective – 'can't get under skin in less than 2 days'– and propaganda 'is really over – now must get down to constructive training … day of teaching so troops do cam. automatically is long overdue'.

No. 2 Camouflage Course was set up in June 1940, again under Colonel Buckley. Stephen Sykes, one of the recruits to this course, remembers how one day in early 1940 he wandered back into the Royal College of Art's studios in Queen's Gate, Kensington from which he had graduated in 1936. He was asked by Barry Hart, one of the tutors, if he would be interested in becoming a camouflage officer, as the War Office had suggested that the College might provide suitable candidates. Eventually, in April, having been accepted into the Royal Engineers, Sykes was interviewed by Colonel Beddington and recruited to join the No. 2 course at the Royal Artillery Camp at Larkhill on Salisbury Plain.

38

Ashley Havinden's thumbnail sketch notes, November 1941.
Imperial War Museum

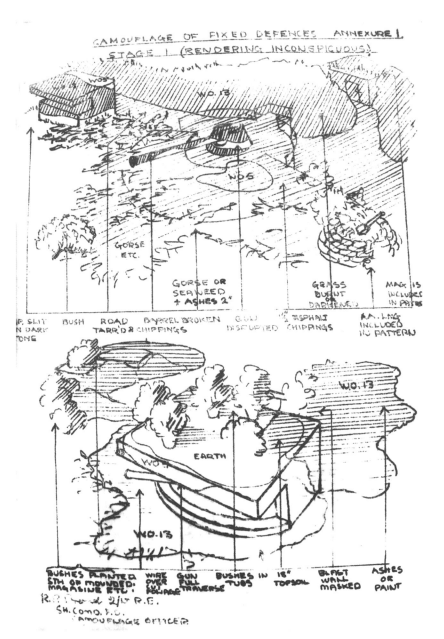

Another recruit to Larkhill was Geoffrey Barkas, cinematographer, who has recorded his camouflage experiences in *The Camouflage Story*.[6] He had been employed as an exhibition organiser by Jack Beddington, the Colonel's brother and publicity chief of Shell Petroleum. He and Stephen Sykes both recall many of the other participants on the No. 2 course, including John Hutton, painter and glass engraver, the sculptor Bainbridge Copnall, and Peter Proud, an art director at Warner Bros Films, all of whom would end up serving as officers in North Africa. Starting with the statutory 10-day military preparation, Barkas's memories of the course include theoretical and scientific studies (light and shadow, colour and tone and their perception from the air); ways of dealing with 'the threat of infra-red photography, wavelengths and the properties of chlorophyll; textures, nets and garnish' and how to conceal pits dug in the ground. After this series of lectures and practical work, the course moved to the Denham film studios outside London where the class was instructed in scenic film techniques such as plaster moulding and other three-dimensional methods (which proved invaluable, according to Sykes, during the period after Dunkirk when a hastily-arranged programme of pillbox construction and concealment was put into action). The team spent the final week back at Larkhill where the course was completed.

Typographer John Lewis, later a long-standing and respected lecturer at the RCA, was also posted to take part on No. 2 Camouflage Course. He describes the teaching as an experimental 'self-help' course to find

Blueprint drawings from a series by Peter Proud entitled 'Camouflage of Fixed Defences'.
The National Archives

out whether flat or disruptive painting techniques were more effective. Lewis notes that after investigation using infra-red photography from the air, it was concluded that the concealment of tracks and troop movements was far more important for the Army than relying on painting and garnished netting to offer protection against air reconnaissance.[7]

Camouflage training was preceded by a short period of instruction in 'the art of soldiering, including map-reading' at Shorncliffe Barracks. Lewis remembers fellow recruits such as the couturier Victor Stiebel, Donald Fraser the opera singer, and Horace Buttery, a picture restorer.

Although the mood was sombre, with the Battle for France over and the exodus from Dunkirk taking place, Barkas records great camaraderie on No. 2 Camouflage Course and many lifelong friendships were struck up. John Lewis's memoirs also bear this out. David Medd, who was posted to Norwich to work under the surrealist painter Roland Penrose, endorses this feeling – camouflage attracted lively-minded designers and artists, and Medd's future career was shaped around the thinking and personalities of some of these individuals.[8]

Roland Penrose lectured to the Home Guard all over England and Wales for two years, preparing them for 'the invasion that never came'.[9] While lecturing at Osterley one day in the summer of 1943 (he was by then an enlisted soldier in the 19th Battalion, Home Guard Middlesex Regiment) he noted that an army staff car had drawn up within earshot of his lecture. The occupant turned out to be General Sir Bernard Law Montgomery, who sealed his approval of Penrose's lectureship by immediately conferring upon him the rank of captain. Soon afterwards Penrose was put in charge of Eastern Command Camouflage School in Norwich.[10] This was a small unit, set up next to the elegant Regency Assembly Rooms, but in spite of its size it was staffed by a charismatic and rather flamboyant group of characters including Oliver Messel, Victor Stiebel, Jasper Maskelyne and Frederick Gore.

David Medd, who had graduated from the Architectural Association in 1940, enlisted in the Army early in the war. He remembers the amazing collection of works by Magritte and Picasso on display in the private quarters of Roland Penrose, who actually had Picasso to stay for a while at the Assembly Rooms at the time of the Paris occupation. Medd was posted to Farnham in 1943, and Frederick Gore became his closest colleague. Although both men were in the development team, part of their brief was to train the troops in 'how to deploy themselves inconspicuously'. Medd's interest lay not so much in the commonly perceived camouflage of 'funny paint shapes' to confuse and conceal, but instead veered towards deceptive techniques based on 'strategy, factors, artefacts deployed to mislead and confuse as well as noise – sonic deception, and light deception'. Although taking on a certain amount of training duties, members of the development team were not given any instruction in specialist techniques, but like the majority of the instructors were picked because they possessed a mixture of imagination and practical common sense.

Oliver Messel was evidently housed in a separate unit near the Cathedral and employed a manservant called Godfrey Winn (who later became one of Fleet Street's most popular gossip columnists). Messel, on the obvious strength of his skill in theatrical design, was primarily employed to devise and supervise camouflage schemes for the east coast defence posts (affectionately known as 'pillboxes') and created many imaginative and memorable disguises, one of which was 'in the appropriately named village of Fakenham'! (The Camouflage Sub-Committee was however, not amused that 'certain specimens of the camouflage of this kind of building were absurd'.)[11] Julian Trevelyan describes the 'pillbox age' as a golden era when unlimited funds were available for indulging designers' wildest fancies. It seems that well after any threat of invasion had passed, Eastern Command continued to build pillboxes for Messel to convert into 'haystacks, ornamental fountains and Greek pavilions'![12] The colourful designer also took any opportunity to decorate the Assembly Rooms 'as if for a Covent Garden show', and in fact the Camouflage Unit's existence was responsible for saving the Rooms for the city.[13]

James Gardner, an exhibition and advertising designer in peacetime, was by then head of the Development Section at Farnham which included Victor Stiebel, Blair Hughes-Stanton and the architect Stirrat Johnson-Marshall. Medd recalls Gardner as 'quirky, fertile and an amazing draughtsman ... very good at getting things done', but he also says that the development team actually had little to do with him. Its

41

work was to undertake the design, manufacture, trial and testing of new equipment for decoy and deception, and was largely involved with creating acoustic and visual equipment such as 'lights, loudspeakers and dummy devices'. David Medd was eventually put in charge of the workshop, developing prototypes which were then sent for production to major companies who would reproduce the designs in quantity.

Another designer who was posted to Farnham later in the war was John Lewis. Following his training at No. 2 Camouflage Course, Lewis had been in charge of the Camouflage School for South-Eastern Command in

Oliver Messel in officer's uniform, 1943.
Photograph by Lee Miller

42

Tunbridge Wells, which was operational from April 1941 until September 1942. It was situated in the new Town Hall, and one of its most popular aspects was that the mess sergeant posted there had been a chef at the Savoy Hotel, thus ensuring that the School was always remembered for its cuisine! The curriculum (delivered by both resident and visiting staff) was supported by frequent aerial reconnaissance expeditions, mainly in reliable old Avro Ansons, as well as Hawker Henleys and Lysanders.

As well as army activities at Tunbridge Wells, John Lewis organised and ran a week-long course for the Royal Navy, which was attended by 20 naval officers and two Marine Officers (John Nash, painter brother of Paul Nash, and the sculptor Alan Durst). There appears to be little documentation of any formal training for naval camouflage. In fact, in January 1943, lecture notes taken by Ashley Havinden ask, 'What is Cam. Directorate Naval Section & shall we have to work with them?' The notes mention that the Ministry of Home Security had a Directorate of Camouflage, with a Naval Section to advise the Admiralty. 'We could do it all for them – & it is quite conceivable that we might have to liaise e.g. over plan to deal with troops leaving this country.'[14]

Lewis's posting at Tunbridge Wells led to a job (through Victor Stiebel, by now in charge of Camouflage Postings) as head of the Canadian Army Camouflage School. After a year, with the course successfully up and running, Lewis returned to London. He was eventually posted to HQ 8th Army at Bari, Southern Italy, where he met his old friend

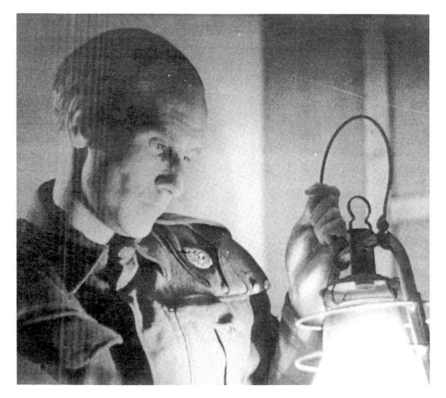

Portrait of a Farnham camoufleur, thought to be James Gardner.
Imperial War Museum

Edward Ardizzone, who was there as an official war artist, and Brian Robb, graphic designer and illustrator, who had been one of the masterminds behind the camouflage and deception campaign which had helped to secure victory at El Alamein. Both of these men were to become RCA tutors after the war.

On his return to Farnham, John Lewis became assistant instructor under Talbot Kelly, a naturalist and a charismatic lecturer on military history. After some months Kelly retired and Lewis took on his role as chief instructor. The Centre was now installed at Pierrepoint House near Frensham. James Gardner was in place as director of development, assisted by Oliver Heath Robinson, son of the famous illustrator, and the architect Basil Spence. Spence and Lewis became lasting friends.

Northern Command Camouflage School, where the

designer and calligrapher William Maving Gardner was in charge, was set up in Darlington, Yorkshire, with practical exercises based on Barnby Moor. Gardner graduated from the RCA in 1939 and was on a travelling scholarship to Scandinavia when war was declared. He returned to volunteer for the Middlesex Regiment from which, on the strength of his design qualifications, he was posted for camouflage training at Farnham. This was followed by a tour of duty in South Midland District until his posting to Yorkshire where 'courses were combined with help for Coastal Command'.[15] An earlier RCA graduate and also a lecturer (later becoming responsible for all personnel) was Eric Turner, and he too remembers assisting Coastal Command, particularly in the camouflage of radio location posts.[16] The diary of David Cooper, who was sent to train at Darlington, records 'a large Victorian Gothic house [called Uplands] in its own grounds: life taken easily and work looks very interesting. Am to mooch around picking up rudiments'.[17] Cooper records that he attended a two-day NCO (non-commissioned officer) course which was 'exceedingly interesting. It's a pity more CO's do not realise the value and possibilities of camouflage. Practical demonstration left one uncertain what is real and what is fake, and imagining hidden men all over the place. I could not have believed that one could hide so many men so completely and utterly in an apparently open field space.'

William Maving Gardner subsequently took over the post of running the Scottish Command Camouflage School in Edinburgh, using a well-known sculptor's studio as premises for delivering courses to 'many units including the Polish Armoured Forces'. Training was carried out with the use of model landscapes, viewed by three groups from different angles: lying down and reporting findings from ground level; cycling round the model to emulate the view seen by a low-flying plane; and taking photographs from ceiling height. Intelligence gained from these exercises was compared and examined for accuracy. Gardner's final posting was to Tunbridge Wells to take over John Lewis's previous duties at Southern Command.

After his training at Larkhill Geoffrey Barkas was commissioned as a first lieutenant, and sent to HQ British Troops, Northern Ireland, as general staff officer, camouflage specialist. His job was to report on the state of existing camouflage arrangements, and to teach the troops methods of concealment and disguise. He and his staff lieutenant set up a programme of military advertising which included training pamphlets, lectures and demonstrations, chiefly in order to improve the standard and understanding of camouflage techniques for army vehicle drivers. The campaign and the subsequent demonstrations were a great success, instilling a practical understanding of the subject in an army of men with little training who had been hurriedly recruited from a civilian background. After four months, however, at the end of October 1940 Barkas received a request from Colonel Beddington to prepare to set sail for Special Operations in the Middle East.

Many of the Royal Engineers officers who had received training on Camouflage Courses Nos. 1 and,

43

THE SAD STORY OF GEORGE NATHANIEL GLOVER

Driver George Nathanlel Glover
Scorned the use of Natural Cover
And never, never could be made
To Park his Lorry in The Shade.
In fact his favourite parking places
Were Vast and Treeless Open Spaces.

When any of his pals demurred,
George Gave them all a frightful bird.
Descending from the dirver's seat,
And using words one can't repeat,
He'd broadcast to the world at large
His Curious Views on Camouflage.

He would remark, "Cor strike me pink,
You must be barmy if you think
There's any need to hide this bus—
The whole idea's preposterous.
Listen, you windy lot of Slobs—
You see these Greeny-browny Blobs?
Well, that's a special kind of Paint
That makes things look like What They Ain't.
No fooling, it's the latest thing—
It's called Disruptured Patterning.
I'm telling you it's pretty hot,
But if you think that's all I've got—
Blimey! You ain't seen nothing yet.
Look here! I've got this blinking NET!"

The net he spoke of was a thing
Composed of neatly knotted string
Which he had thrown across the bonnet,
With not a stitch of garnish on it.

"So what with this here paint and net,"
Said George, "I'm open for to bet
My bus is almost if not quite
Invisible to Human Sight.
Let them as wants to run and hide.
I'll stay right here. I'm satisfied."

His comrades said, "Excuse us, please,"
And sought the cover of the trees—
An act of Sterlilng Common Sense
And well-informed intelligence,
For George had hardly said these words
When German planes arrived In Herds
And landed several tons of Muck
Right on the top of Goerge's Truck.

Emerging later, safe and sound,
His comrades searched for miles around
But Not One Trace did they discover
Of Driver George Nathaniel Glover.
And I am also very sorry
To say they never found his Lorry.

The moral is, as all can see,
Ungarnished Nets are N. B. G.

'The Sad Story of George Nathaniel Glover'.
Advice for British troops by Geoffrey Barkas, The Camouflage Story, Cassell plc, a division
of the Orion Publishing Group (London), 1952

44

mainly, 2 were billeted to Egypt with the 8th Army. Geoffrey Barkas, now elevated to the rank of colonel, was posted to oversee camouflage operations in GHQ Cairo. After having spent a year working on camouflage schemes which were totally different in character from anything experienced at home, and often in desert situations, he was credited with setting up the Camouflage Training and Development Centre at Helwan, near Alexandria.[18]

The Centre was situated in a camp of around 60 huts, allowing 20 staff and 150 students at a time to work there. Ashley Havinden records, in notes from a CD&TC lecture by Julian Trevelyan on 9 June 1942, 'big camp … austere – 6 huts on edge of camp … 3 big lecture rooms'.[19] Adjoining the base depot was a 'patch of desert' containing an exceptional variety of different types of desert terrain – this was to be used as a development ground for testing camouflage techniques and prototypes. This area would also be used for demonstrations and practical exercises.

John Morton, a corporal in the Searchlight Regiment with a cabinetmaker's training (later to collaborate with David Pye, architect, and Robert Goodden, an architect and designer, on major projects for the Festival of Britain) was posted to Egypt to be involved in field-work as a carpenter in the Suez Canal area. He noticed one day that GHQ were advertising for troops interested in camouflage work, and applied. He was interviewed by Geoffrey Barkas, whom he describes as gentle and unmilitary (his initial greeting was, 'How nice, won't you sit down?'), and consequently

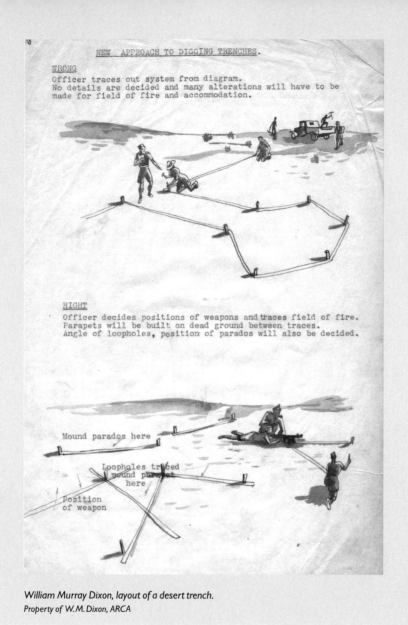

NEW APPROACH TO DIGGING TRENCHES.

WRONG
Officer traces out system from diagram.
No details are decided and many alterations will have to be
made for field of fire and accommodation.

RIGHT
Officer decides positions of weapons and traces field of fire.
Parapets will be built on dead ground between traces.
Angle of loopholes, position of parados will also be decided.

Mound parados here

Loopholes traced
mound parapet
here

Position
of weapon

William Murray Dixon, layout of a desert trench.
Property of W. M. Dixon, ARCA

45

Hugh Cott, the animal-colouration expert from Farnham, became chief instructor at Helwan. In addition to his specialist knowledge and his teaching expertise, he had by now accumulated several months' experience of camouflage in the Western Desert. There were three main camouflage courses. One was for senior officers (to teach them about 'putting over camouflage'), another was for training instructors who would travel with mobile units and help with small jobs, and a third was for syndicates with sergeant instructors who had to go out into the desert to help troops learn the best methods of concealment.[19]

The staff officer responsible for training and development was the art director Peter Proud, recently returned from Tobruk where he had built up an active and successful camouflage force of around 300 men. According to another ex-RCA student, William Murray Dixon, Captain Proud had been sent out to Egypt to 'find out how camouflage should be developed in the Middle East, and do it'. He consequently contacted the War Office to find out if there were any RCA graduates available locally for recruitment, and Dixon was introduced to him.

The imagination and resourcefulness of all these young artists and designers when they went into war as camouflage officers, must have made learning great fun for troops who were subsequently under their command. None of the officers had set out to be instructors, yet their lively inventive minds and their practical skills allowed them to become natural teachers by way of concept, demonstration and inspiration.

transferred to the Royal Engineers with the rank of sergeant in a camouflage unit based in Cairo. When his unit moved, Morton was sent to Helwan for a month's training at the Centre, and was subsequently given the job of lecturing to troops on desert camouflage.

CHAPTER FOUR

Civil Camouflage

IN 1940 THE UNITED KINGDOM was subjected to a relentless series of air-raids, and as a result of their devastating effects they served to raise public awareness that the camouflage of recognisable landmarks was of fundamental importance. The Ministry of Home

Security's plan to camouflage key industrial sites was put into action, with schemes intended to 'confuse a pilot at a minimum of 3 miles distant and 5,000 feet up during daylight'. Schemes did not need to be specifically designed for night attack because enemy recognition would be impaired by darkness. In addition, although the visible natural features of the British Isles were too numerous to disguise, it was decided that camouflage treatment should be arranged if a particular landmark was an obvious guide to a potential major target.

The Ministry of Home Security recommended in the same year that 'there is no necessity for any secrecy in the subject ... an intensive educational campaign would do much to make the public understand why and how camouflage is used'. Before the public became aware of the programme in place to assess the vulnerability of potential targets, local pressure to camouflage unimportant buildings grew steadily, thus forcing the country into the unnecessary expenditure of a large amount of money, materials and labour in order to camouflage buildings which were not considered prime targets. Meanwhile, blackout regulations

Opposite: T. E. La Dell, The Camouflage Workshop, Leamington Spa, 1940. Officers and technical assistants at work on assorted models. Foreground, from left to right, Thomas Rathmell, Captain Lancelot Glasson, Captain Gilbert Solomon, and Christopher Ironside. Centre, Betty Fiddeman. Imperial War Museum

Left: Eric Ravilious, sketch of a barrage balloon, possibly for Barrage Balloons at Sea, 1940. Private Collection

48

were implemented at the earliest possible stage, and the public willingly took these on as its responsibility towards the war effort. In October 1940 the Home Office commented that 'the Blackout is the essential background against which all forms of concealment and deception at night must be set'.[1]

In an excerpt from Robin Darwin's 'notes' appraising the role of artists in camouflage, sent to Mrs M. W. Murdoch, Public Relations Officer to the Ministry of Home Security on 18 February 1943, he states that although camouflage can never entirely fool the camera, 'the bomb aimer must rely on what he sees with his eyes and a moment's doubt, the slightest hesitation may send his bomb far wide of the mark'. If camouflage could momentarily postpone identification of a target, a few extra seconds might enable guns to be brought into action or fighters to get on the tail of the raider. It might also 'give time for the workers in a factory to reach their shelters and so save many lives. At night when camouflage can still play a most important part', the bomber, while searching for his target, could be lured into 'the dangers of the balloon barrage' and thus brought down.[2]

Work began in prioritising the key sites requiring concealment, and the camouflage team increased its numbers in order to organise and design schemes for the protection of these vital plants. On the CDCE's transfer to Leamington in 1940, Captain Glasson recruited heavily from the fields of art and design, as well as bringing in scientists and essential Forces personnel. Newly appointed camoufleurs included, as

CE/8/C/10A/6.

Camouflage Report on Water Tower at Chew Magna.
----

1.   This tower which is built of concrete and measures approx. 30' in diameter by 30' in height is sited at the corner of a field and has a good background of trees of almost equal height.

2.   It could be made unobtrusive by darkening down the concrete or, if it was regarded as sufficiently important, by applying a bold disruptive pattern.

3.   The present method of concealment is ineffective.

4.   It consists of a main coat of light grey paint (little darker than the original concrete) with a weak design of light Olive green added. There has also been a small amount of disruption attempted with a black paint, but the principle has been misunderstood and this is of no value.

5.   Presumably no anti alkaline primer or waste oil has been used as a sealer with the result that the paint is "chalking" badly and is already flaking off.

J.O. Smithson Lt.R.E
for .Captain, R.E.
G.S.O.3 (Cam.).

H.Q. 8 Corps,
Home Forces.
23 Apr 41.
FGB/DJ.

Extract from a report by Julian Trevelyan highlighting misconceptions about the industrial use of camouflage, and giving his practical suggestions to rectify this issue.

well as the relocated London team (who, according to Julian Trevelyan, grumbled 'incessantly about the monotony of life in Leamington and about their chief, Captain Glasson'), painters Mary Adshead, Rodney Burn and Colin Moss.[3] The journal *Nature*, in its 22 June 1940 issue, records that 'of some sixty-five technical officers, all but four are either professional artists or, at the time of recruitment, were students at art schools'. This rather scathing article, while correctly supporting the need for biologists as members of a camouflage team, deplores the employment of artists who lack the 'fundamental biological and psychological principles involved'. A riposte from C. H. Rowe in the issue dated 3 August points out that the artist is 'no mean naturalist himself', having learned the principles of protective colouring as understood by Leonardo da Vinci and taught in art schools since the time of Rembrandt.

Richard (Dick) Guyatt was an early member of the Leamington team, having moved there when the headquarters relocated from Holborn. As a graphic designer he had been teaching part-time at Hackney School of Art when his principal Cosmo Clark, instrumental in establishing the Civil Camouflage Section, suggested in early 1940 that Guyatt apply for a post. Both Guyatt and the designer and teacher Christopher Ironside were successful in their interviews, and were employed on salaries of £400 a year.[4]

Owing to the widespread nature of camouflage activities it became necessary to create a new rank of officer. A number of regional camouflage officers were appointed to keep the executives of the different Services in touch with all camouflage arrangements and the realisation and maintenance of schemes. Each RCO worked closely with the appropriate technical officer from the Construction Section. Dick Guyatt became regional camouflage officer for Scotland, with Leamington remaining his base. He and the rest of the team were constantly involved in the development of camouflage schemes for essential buildings and services. Julian Trevelyan remarks that man's 'more extraordinary activities especially in time of war, are those that are often the most conspicuous. The shine on large areas of roof, the excessive regularity of the typical depot … seen against the irregular background of the countryside, the plumes of smoke that rise from so many industrial activities, these are some of the "give-aways"'[5] The skill of the camoufleurs was required to conceal or direct unwelcome attention away from sites such as these.

Early in the war, before the introduction of the camouflage department, some appalling examples (how *not* to camouflage) had been observed. Examples of these, according to Hugh Casson, were 'magnified [stylised] trees … painted in elevation upon the lofty sides of some concrete cooling-towers'; an urban cinema roof in 'alternate stripes and patches of grey, light yellow and a delicate apple green'; and the 'dazzle-painted monuments to the gullibility of Big Business which … could be seen in the suburbs of our big cities'.[6] *Nature* (June 1940) refers to double-decker bus roofs painted in grey or green and brown, the sides

49

50

Colin Moss, Power Station, purchased in 1943. The suburban power station at Stonebridge Park has been camouflaged using a mixture of techniques, yet appears more, rather than less, obvious! In the foreground, water has been camouflaged by means of suspended netting.
Imperial War Museum

remaining red – an instant target for an attacking plane. In the words of typographer-turned-camoufleur John Lewis, 'whoever was responsible for civil camouflage had instigated a programme of painting green and brown stripes on factory buildings, so that they were the only noticeable buildings in industrial wastelands'.[7] In spite of the fact that camouflage was intended to conceal or disguise, in these cases it meant 'to make noticeable' – like 'a Macgregor tartan at a Brixton funeral. Maybe good advertising – but good advertising is the antithesis of camouflage'.[8]

Although these examples were ideal 'anti-artist' material for the 'biology' lobbyists, it seems, in fact, that the bigger paint companies were often to blame. Bodies such as the Paint Manufacturers' Association would encourage their members to commission their own unauthorised and inappropriate designs and carry them out for advertising purposes. Julian Trevelyan, in a letter to the *Observer* (14 January 1940) notices 'the very amateurish jobs that one sees along the By-passes. Example, a cinema in the Kingston By-pass which has only camouflaged its roof'. Once the CDCE had been firmly established, the Ministry of Home Security was quickly able to put a stop to these aberrations which often dangerously served to draw attention to a site, rather than away from it.

Official visits would be made to assess camouflage schemes. A series of letters from the Ministry to Dr Stradling and Professor Curtis, written in September and October 1940, record arrangements made for the chairman and members of the Select Committee on National Expenditure to visit the Camouflage Establishment, to see camouflaged buildings and 'the Austin Motor Works at Birmingham from the air'.[9] This particular visit was rather disappointing because by the time arrangements had been confirmed, the possibility of a flight was complicated owing to necessary fighter activities. However, the day was considered a success, particularly as the important matters of lunch (in Leamington) and 'a cup of tea in the afternoon' were duly scheduled into the timetable! The result of the inspection was that five recommendations were made, including one that 'industrial undertakings' requiring camouflage should be obliged to consult the central camouflage organisation. Instead of the major paint companies cashing in on the chance to advertise, paint formulae and colours would need to

The Leamington camouflage team at a local airfield. Fourth from left is Christopher Ironside with, on his right, Eric Schilsky who later married Victorine Foot.

Property of Felicity Fisher

comply with the work being executed at the Paint Research Station.

The most important aspect of the camoufleur's job involved viewing the installation or site in question from the air. Richard Guyatt remembers the RAF flying unit at Baginton, which was attached to the Leamington headquarters. The RAF Photo Unit was based at the Regent Hotel. It seems that pilots who were past the age for 'operational work' were the ideal guides for designers on camouflage survey missions. The Static Camouflage Assessment Committee would consider the importance of a site, the probability, probable nature and consequences of attack, and the general character of a target

for the assessment of its treatment. Once aerial photos had been taken, the 'precise interpretation of necessary measures must [then] be left to the Camouflage Officer'.[10] The term 'static camouflage' refers to all historical, official, commercial and residential buildings, supply depots and installations, also to all non-mobile accommodation and equipment used by the Forces. In the case of airfields this would include factories, depots, hangars and other related buildings.

One of Richard Guyatt's lasting memories is that in 1941 a 'new boy' arrived at the Directorate, one of a small group of recruits including the painter Stephen Bone, who were to join Captain Glasson's existing

General responsibility for all Regions - Captain G. B. Solomon.

Region	Camouflage Officer	Assisted by:	Technical Officer	Assisted by:
1 & 2	Mr. H.G.Hoyland	Mr.C.E.J.Shelley	Mr.D.W.L.Daniels	Mr.R.Dobson,Mr.J.Leech
3 & 4	Mr. P.B.Hickling	Mr. F.S.Eastman	Mr.F.W.C.Adkins	Mr. A.Millington)
				Mr. A.L.Osborne)
5	Mr. C. Ironside	Mr. W.E.C.Morgan	Mr.G.L.M.Jenkins	Mr. E.G.Dean.
6 & 12	Mr. S. Bone	Mr. G.Grayston	Mr. G.L.M.Jenkins	Mr.,E.G.Dean.
7 & 8	Mr. E.S.Drake	Mr. G. Watson	Mr. C.R.Disney	
9	Mr.L.J.Stroudley	Mr. G.S.W.Malet	Mr. F.W.C.Adkins	Mr. A.Millington)
				Mr. A.L.Osborne)
10	Mr.H.G.W.Irwin	Mr. L.C.Duffy	Mr. D.W.L.Daniels	Mr. R.Dobson,Mr.J.Leech
11	Mr.R.G.T.Guyatt	Mr. R.Darwin	Mr. D.W.L.Daniels	Mr. R.Dobson,Mr.J.Leech

Mr. W.T.Monnington will be responsible for the inspection of all
Civil Aerodromes for which C.D.C.E. are responsible, together with
all associated buildings (factory or R.A.F.). He will prepare and
maintain schemes of camouflage for these, in co-operation with the
camouflage officer originally responsible for the factory (if any).

The foregoing arrangement will come into force as from 1st December,1940.

Leamington staff list for December 1940.

team. This was Robin Darwin, who soon found that he was to be on Guyatt's staff. For some reason this idea did not appeal to him, and magically he managed almost immediately to secure a new post as secretary to the Camouflage Committee in London, a position of much more suitable stature (according to Guyatt) and one in which he was to show a great deal of compassion to the sometimes misunderstood camouflage artist or designer.

Robin Darwin's paper 'Artists in Camouflage' contains a description of the routine way in which a camouflage officer would fly over the target in question, taking notes and photographs from every angle. The officer's artistic training would enable him to see how the target's appearance could be changed, and at the same time his experience in camouflage would allow him to effect these changes.[11] A scheme would then be developed, 'from the first rough sketches to the

Opposite: Robin Darwin, Camouflaging the New Flight Shed, 1941. This was probably painted before Darwin became secretary to the Directorate, perhaps at Baginton airfield. Imperial War Museum

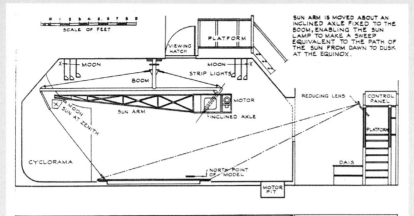

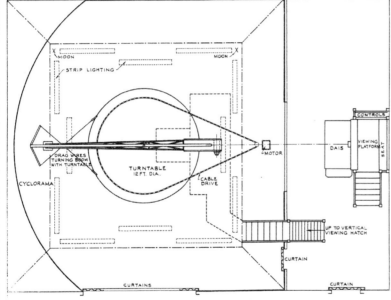

SUN ARM IS MOVED ABOUT AN
INCLINED AXLE FIXED TO THE
BOOM, ENABLING THE SUN
LAMP TO MAKE A SWEEP
EQUIVALENT TO THE PATH OF
THE SUN FROM DAWN TO DUSK
AT THE EQUINOX.

Above: Diagram showing how the 'viewing room' at Leamington operated.

Right: Christopher Ironside (left) at work on a model, probably with 'Johnnie' Walker.
Property of Virginia Ironside

finished perspective drawing', which could be handed
to the contractor commissioned to execute the task.
Detailed coloured drawings were issued for the
simpler schemes; in cases where a single colour could
be used to simply 'tone down' an installation, verbal
specifications were considered adequate.

For crucial targets, or where the scheme required a
particularly detailed design, it was considered necessary to create a scale model of the site and its
surroundings. Christopher Ironside became deputy
senior design officer, and was put in control of all
modelmaking. A special chamber at Leamington HQ
was allocated as a viewing room similar to the one at

the RAF's London headquarters at Adastral House, with a turntable where models were placed and assessed from every angle under simulated atmospheric and lighting conditions. Darwin describes how 'all different kinds of light from diffused light to bright sunlight and from starlight to moonlight' could be accurately imitated and controlled. 'The "Sun" swings round on a great arm and can be fixed at any altitude; or it can revolve in relation to the turntable, so that the designer standing … at a distance from the scale model repre-

senting approximately five miles, can imagine himself in an aircraft circling round the target', and so would be able to assess the effect of his camouflage scheme from every viewpoint.

Another description of the viewing room taken from *The Fortnightly DO* records that in 'a vast landscape set out on the ground, in front of a sort of painted cyclorama' models 'can be seen through viewers, which, by turning a handle bring the image from a little pinpoint in the distance nearer and nearer just as a bomber

Anne Newland, The Rink, c. 1942. An ink and wash sketch of the studio interior, showing the internal 'viewing room', and the 'sun' as described by Robin Darwin. RAF Museum

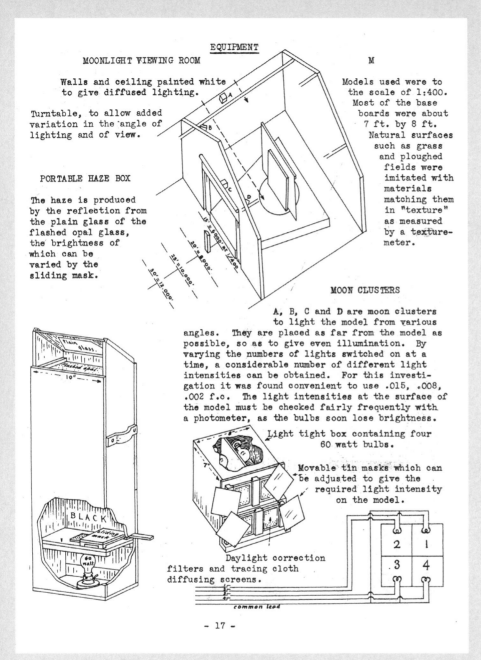

Diagram of the 'moonlight viewing room', with the 'haze box' and 'moon clusters' used to create different atmospheric and lighting conditions.

would see it; which would suggest that civil camouflage officers don't need to fly. But this is not the case, as the air around Leamington is thick with them.'[12]

Although most air-reconnaissance surveys were made by day, the growing frequency of air raids under cover of darkness meant that night conditions had to be fully understood. It was imperative that schemes were designed to conceal installations in moonlit conditions as well as by day. A scale-drawing of the 'moonlight viewing room' at Leamington shows the room with walls and ceiling painted white, and different methods of adjusting lighting including a periscope-like construction called a 'portable haze box', used to re-create less than clear conditions.[13]

The Ministry of Home Security's Research and Experiments Department (R&E) carried out a detailed investigation into the camouflage of factories in moonlight conditions, helped by 'experienced officers of the Directorate of Camouflage', in order to assess the changes from day to night vision and how the observation of likely targets could be affected.[14] Tests were made by viewing charts and models under controlled conditions, supplemented by observational night flights up to a height limit of 7,000 feet. Resulting reports showed that visibility distance depends on atmospheric conditions and the tonal contrast between objects and their surroundings. Findings indicated that many factories were conspicuous at 7,000 feet in varying moonlight conditions, usually being revealed by pale features such as roadways and faded roofs. Factory walls, being particularly long and straight, made very

PRINCIPLES OF NIGHT VISION

An observer at night must make allowances for an actual physical change in the working of the eye when lighting is reduced to starlight levels of illumination. Even by moonlight this change is beginning to take place.

The change is due to the retina having two types of nerve-ending, CONES, which react to bright light, and RODS, which come into their own when it is nearly dark. The change-over to NIGHT VISION has the following effects :—

AT LEAST 20 MINUTES IS NEEDED FOR EFFECTIVE DARK ADAPTATION

If temporarily dazzled by a bright light, the eyes will soon recover dark-adaptation except that they may suffer from " after-image." Red light will not spoil dark-adaptation.

LIGHT OBJECTS STAND OUT RELATIVELY MORE BY NIGHT THAN BY DAY

The darker a tone is, the nearer it approaches the level of brightness below which the eye cannot see at all (called the " threshold " of vision). Differences between black and a " middle " tone are therefore more difficult to distinguish than between a " middle " tone and white, which makes light tones seem relatively lighter by night.

2

THE SENSE OF COLOUR IS LOST, AND THE TONE OF SOME COLOURS ALTERS

The " rods " are insensitive to colour. Dull colours fade to grey, bright red looks black, blue looks pale grey. This rule does not of course apply to coloured lights whose brightness is above the level at which this effect takes place.

POWER OF PICKING OUT DETAIL IS REDUCED

This is due to the sparse distribution of nerve fibres serving the " rods " and is especially noticeable with dark objects where the eye's power of perceiving contrast is less. The lines of shadow in a factory roof will often blend together from quite a low altitude, except sometimes when the contrast is increased by the lighted parts of the roof reflecting the moon. Very roughly, the eye's power of " resolving " detail is five times less by the brightest moonlight than by day.

VISION IS CLEARER OUT OF THE CORNER OF THE EYE THAN STRAIGHT AHEAD

This is because there are no " rods " immediately opposite the front of the eye, so that the sides of the eye are more sensitive by low levels of lighting. The habit of not looking straight at objects which attract attention by night should be cultivated.

3

(C47636)

B 2

'Principles of Night Vision' from Notes on Aerial Observation at Night, *one of the many official handbooks written for camouflage officers.*

obvious shadows, as did angled 'sawtooth' roofs, although these were not always conspicuous unless shiny. Other man-made features easily seen in moon-light conditions were plumes of steam and smoke, scarred ground, road junctions, railway embankments and built-up areas in general. Dark areas of woodland and the sheen on a river's surface could also lead to the identification of a crucial site.

Science played a very important part in these invest-igations. The eye's sense of colour is greatly diminished

in moonlight situations, and it is five times less able to resolve detail than in daylight. Hugh Casson, responsible for airfield camouflage, writes about the importance of the camoufleur's familiarity with the basic principles of light and colour, and the effect of tone and texture.[15] Tone is more important in camouflage than colour (in night conditions landscape is 'seen in monochrome' and colour has no value), but general tone from the air appears much darker than expected owing to integral shadows. However, a palette of camouflage colours – black, greens, browns and greys in keeping with the British landscape – was developed in accordance with the evidence gathered from aerial surveys. Texture could be used to destroy smooth light-reflecting surfaces and disperse light in all directions, therefore creating a mid-to-dark surface. Form, revealed by shadows cast by the planes composing the whole, could be concealed by texturising horizontal surfaces, and by painting light areas dark and dark areas light.

Casson makes an interesting point, again in his *Architectural Review* article, that the preferred building style immediately before the war (particularly for industrial installations) was the angular, pale, stark modernist concept, in which buildings were 'designed and placed to contrast rather than merge with their surroundings' and often constructed with hard, light-reflecting materials. The shortages and restrictions of war only served to encourage this approach, increasing the number of buildings which were harder to camouflage than those more organically in tune with

Opposite: Hugh Casson, watercolour of airfield buildings, 1943. The sheds have been camouflaged using a disruptive design. © Hugh Casson. Reproduced with permission of Sir Hugh Casson Ltd

their environment, and this tested the ingenuity of the camoufleurs.

A set of basic camouflage techniques suitable for day and night conditions was established, and it was the job of the Leamington team to use its inventiveness and imagination in developing individual schemes. A booklet entitled *Cheap Methods of Darkening Buildings* portrays many resourceful methods designed to use wartime waste materials.[16] The siting and planning of new buildings was of great importance, but as most installations were already in place it was necessary to devise the most appropriate scheme for each one. Roads and concrete surfaces needed to be toned down with a mid-to-dark colour other than black, and texturing had to be used to suppress shine and also to imitate natural surfaces such as grass, ploughed earth, woods and dense vegetation. Conversely, fake roads were painted or created with netting, often right across factory roofs in order to draw attention away from the outline of the building. Screening ('fadeout') to delete tell-tale shadows, with netting, scrim or steel wool, was expensive although as a rule only two-thirds of a surface needed to be covered: this was achieved by means of lean-to structures constructed to support the material. Patterned camouflage proved to be effective by moonlight but needed to be broader in scale than that successfully used in daylight; black and mid-tones worked best near small areas of woodland. Glass factory roofs required special treatment to dull or eliminate reflection. In addition to the outline and components of a factory building, instructions were

59

60

A page from Prevention is better than Cure, *a booklet produced for camouflage staff.*
The Establishment figure could be a cartoonist's impression of Robin Darwin.

T. E. La Dell, Fixing
Textured Netting for
Camouflage, *1943*.
Imperial War Museum

drawn up for its occupants. Car parks had to be camouflaged, or special parking arrangements put into
action; outdoor stores and stocks of materials were an
instant giveaway to the nature of an installation, so
light-coloured stores were covered with a dark tarpau-
lin; meanwhile workers were advised not to wear
light-toned overalls as these were clearly visible from
the air.

Areas of water, very easily distinguishable from the
air by day or moonlight, needed special treatment.
Various methods were used, including rafts covered in
camouflage materials and supported by tin drums or
willow structures, or 'floating dust', particles suspended in oil but effective only if the water was still.
A more radical approach involved the drainage of lakes
or ponds, and although the resulting mud echoed the

62

original shape, this could be concealed by man-made cover or longer-term horticultural treatment.

As well as special paints in made-to-order colours, waste materials played an enormous part in the job of disguising the landscape. Camouflage was a natural consumer in the wartime ethic of 'waste not, want not' and much industrial refuse was recycled in the effort to conceal roads, buildings and scarred ground. The resourcefulness of the camoufleurs meant that mater-ials frequently used included coal tar and its deriva-tions, waste vehicle oils and sludge, clinkers, cinders and steel slag, coke chippings, cement slurry and gravel. Roads were often disguised with coated or coloured chippings secured with adhesive dressing; factory roofs were obscured with sand, cork granules or cement slurry and gravel; feathered wire netting (supplied ready-painted by the manufacturer, Cullacorts) and steel wool were also used for this

Colin Moss, Water Camouflage, *purchased in 1943, showing how reflection and movement of water have been disguised by netting with central floating supports. Imperial War Museum*

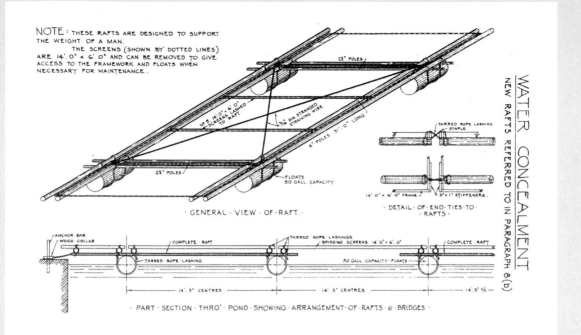

Diagram of a water concealment system which used oil-drum 'floats' with a wooden framework to support the opaque screens.

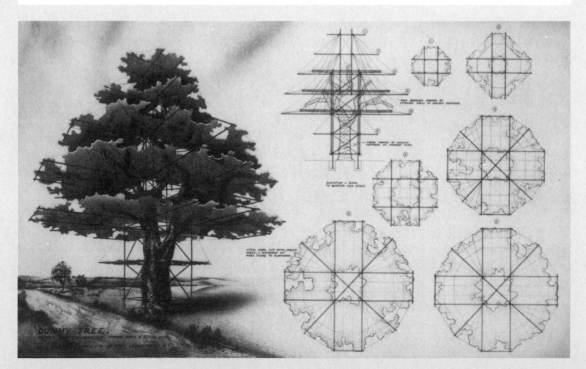

Dummy tree designed by Joseph Gray using 'Cullacorts' netting and steel wool, supported by a metal framework.
Imperial War Museum

purpose. Agricultural waste was also put into action – *Cheap Methods of Darkening Buildings* mentions a combination of coal-tar and cow-dung. A mixture of sifted earth with pigmented powder or plaster of Paris was spread on bare soil to colour it, and clinkers and cinders were also scattered over damaged ground.

Steel wool camouflage material (a different product from the domestic version) was developed early in the war by Major Joseph Gray, an artist in the Camouflage Section dealing with ordnance factories and other army installations. It was a durable covering which could be manufactured in quantities large enough to disguise even the most sizeable installation or landmark. It was far superior to the three-inch netting textured or 'garnished' with scrim, having a more convincing depth to its surface, as well as the added advantage of being non-flammable. Steel wool was therefore used for crucial structures such as oil terminals in addition to radio stations, operational centres and all official headquarters. It appears that there was some confusion as to whether Gray was the inventor of this material, as borne out in a series of remarkable testimonials written in 1947 by key camouflage figures including Honorary Air Commodore Norman Wilkinson, Colonel Frederick Beddington and Lieutenant-Colonel Francis Wyatt, which were sent to the Ministry of Supply in support of his achievements.

A set of Christopher Ironside's very detailed annotated aerial inspection photographs show 'before and after' camouflage schemes, and illustrate the meticulous approach of the camouflage team to its work.[17]

The pictures clearly indicate the hazards caused by changing light direction (these photographs were taken 'up sun' and 'down sun'), the care required when camouflaging in unfavourable conditions, and the dangers of introducing man-made alterations to the landscape such as smoke plumes and scarred ground. 'Before and after' documentation shows successfully completed camouflage schemes. A factory and its surroundings are retouched to darken all conspicuously light areas of the building and its environs; another factory is disguised as a housing estate, and a third imitates an area of woodland. The straight edges of a group of aircraft hangars are distorted by the use of netting; a research station is transformed into a group of houses by building artificial pitched roofs and chimneys onto existing flat roofs. Other images show the perils of conspicuous local features (man-made and natural) such as railways, roads, reservoirs, rivers and canals. An awareness of how installations might appear to a reconnaissance aircraft meant that paramount importance was placed on their siting and method of construction. One photograph shows a near-perfect example of architectural design, which is virtually invisible at first glance, and additional concealment is provided by an artificial hedge snaking across the roof!

Colonel Wyatt's summary of the complete camouflage operation, produced as a record on behalf of the Camouflage Committee of 1943 (which comprised senior camouflage staff from all the Forces), states that camouflage treatments must include conspicuous

Opposite: T. E. La Dell, A Machine for Spraying Scarred Ground for Camouflage, purchased in 1943. A tractor pulls one of the many agricultural or industrial machines adapted for wartime use. Here, coloured matter is discharged to disguise tracks and spoil. Imperial War Museum

APPEARANCE OF THE LANDSCAPE

Reduced illumination and haze interference obliterate a large number of the factors used for recognising features of the landscape by day. In place of clear outlines and a wide range of tones and colours, the landscape even by bright moonlight appears merely a general silvery grey, mottled with a darker tone and a lighter tone, the degree of contrast depending on atmospheric conditions and altitude. If such scanty data are to enable the recognition of an objective, the observer must be well primed beforehand as to what to expect. Landmarks likely to identify the target area can be chosen from the map. Photographs of the area should be studied if possible in an epidiascope with a moonlight filter attachment. Allowance must be made, however, for changes in appearance due to the angle or direction of lighting shown in the photographs differing from what they will be at the time of observation.

The effect of a change in the angle of lighting on, say, a building will be obvious.

But also THE RELATIVE LIGHTNESS AND DARKNESS OF NATURAL SURFACES WILL ALTER WITH A CHANGE IN THE ANGLE OF LIGHT ON ACCOUNT OF DIFFERENCES OF SURFACE TEXTURE.

For instance :—
A *ploughed field* may look lighter than a *grass field* seen from above by a HIGH SUN or MOON.

But by a LOW SUN or MOON, the furrows fill with shadow, while the grass, being translucent, is not darkened to the same extent.

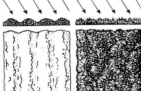
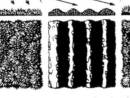

Roads, flattened earth and *sand* look light from above because their light natural tone is not much darkened with contained shadow.

12

TEXTURE ALSO CHANGES THE RELATIVE TONES OF SURFACES WHEN SEEN FROM DIFFERENT ASPECTS. The *ploughed field* looks lighter than the *grass field* when seen down-sun or down-moon, but darker from the opposite aspect where the grassland is translucent and the ploughland is full of shadow.

This effect is particularly important in the case of water.

Still clear water reflects sun or moon with a flash, and looks dark from all other angles (except for sky reflections, but these usually are only bright enough to cause water to appear lighter than land at low angles of view).

The more acute the angle of reflection, the brighter the flash.

Ruffled water scatters the light so that it will not look quite so brilliant at the angle of flash but will be fairly bright over a wider angle.

" TEXTURE " DEPENDS ON THE NATURE AND THE SHAPE OF THE SURFACE

Nature.—Whether the surface disposes of the light which falls upon it by reflecting it (water), by scattering it in all directions (chalk), by absorbing it (soot), or by reflecting it internally as " translucency " (grass). Most surfaces combine more than one quality.

Shape.—Whether it is flat, or so rugged as to produce contained shadow.

13

Training notes for camouflage officers concerning the effects of varying lighting conditions on the attacker's perception of the landscape.

adjacent features.[18] Factory and civil airfields were a case in point: their layout included dependent hangars and other large buildings, so the remit for their concealment fell to the Ministry of Home Security. Richard Guyatt relates how one of his tasks was to organise camouflage for Ringway Airport, Manchester.

As an illustration of the precarious nature of the job, he recalls that during a reconnaissance flight in a Gypsy Moth plane, the propeller 'fell off, taking the front of the plane with it'. The pilot managed to circle the airport for half an hour in the badly damaged plane while he jettisoned the petrol, before making a brilliant and

Cedric Kennedy, Erecting Camouflage Netting at a Factory, *1942, illustrates the very precarious nature
of some of the work undertaken. The supporting poles have also been camouflaged. Cedric Kennedy was a
member of the Leamington camouflage team.*
Imperial War Museum

Reference..................................

Mr. Imrie.
S.A.
C.C.O.

Visit of Mabane Committee.

    The Committee will visit Leamington and Compton Verney on Thursday, 17th April and 18th April, following Easter.

    I should like to discuss detail arrangements on Saturday forenoon.   If that is inconvenient I would do so at 9.30 a.m. at the Rink on Wednesday, 9th April, 1941. Please arrange as you prefer.

    Among the items I am particularly anxious that they should see are:-

        Slides representing good and bad camouflage and before camouflage.

        Models in the Viewing Room with removable nets to show the value of the nets.

        Models in moonlight, flare light, day light and sunshine arranged to show importance of shadow, colour, tone and texture.

        Viewing through a bomb sight.

        Water Camouflage. Rafts & Coal Dust at Compton Verney.

        Full scale samples of nets with scrim, steel wool and if possible B.G.

        These should include the model of our new design: this is very important.

    We depend greatly upon this visit to make the Committee realise the value and effectiveness of Camouflage. The Committee's decision will govern the whole camouflage Policy probably for the remainder of the War. We must therefore give an adequate account of what we can do in really useful Passive Defence.

        (Signed) T.R. Cave-Brown-Cave.

7th April, 1941.

Memo regarding an official visit to Leamington in April 1941 by the Government's Mabane Committee on Concealment and Deception, which was highly influential in camouflage policy. The Committee was chaired by Sir William Mabane, Parliamentary Secretary, Ministry of Home Security.

68

dramatic landing, using one of the wings as a brake. (Camoufleurs had to be not only resourceful, but fearless too!)

The resulting scheme was tested out on a 'vast model the size of a room' and, when approved, transferred full-scale to the airport itself. Later, when Guyatt was carrying out a ground inspection, the airport manager proudly pointed out his own house standing on the perimeter of the airfield. It had been camouflaged with a large, carefully executed dribble of green paint down one of its white walls – faithfully reproducing the careless painting of the model. Guyatt duly admired this, and in his own words, 'sleeping dogs were left lying!'

A paper produced by the Ministry of Home Security entitled 'Defence Measures for Industrial Concerns' mentions the trial 'night decoy of a Northern town', a replica installation designed to lure attackers away from the actual industrial town which was a prime target.[19] Fake blast furnaces emitting smoke, as well as decoy street and factory lighting, were apparently very successful in attracting bombs away from the objective. In retrospect, it was decided that 'searchlights, anti-aircraft guns and a smoke screen over essential parts of the parent town' would have helped even more. The effectiveness of this experiment was to be endorsed and developed for the Air Ministry by the imaginative work of Colonel John Fisher Turner and his day and night decoy airfields, complete with counterfeit planes created by experts from the fantasy world of Sound City Film Studios.

Louis Duffy, Factory Entrance, 1943, shows the use of disruptive patterning in an angular as well as an organic configuration.
Imperial War Museum

CHAPTER FIVE

RAF Camouflage

THE OMINOUS EXPANSION of Germany's fighter plane division, the dreaded Luftwaffe, in the mid-1930s had the effect of prompting the British Government to urgently consider the state of its military air bases. Up until 1935 there were fewer than 50 RAF airfields, some of them antiquated survivors from the First World War, others in various states of modernisation. Government rearmament policy resulted in the allocation and construction of many new airfields; by the time war was declared, twice the original number were in operation.[1] By 1945, at the end of the war, Britain's aerial defence development was so expansive that the total number of military airfields had grown to 740.

The Royal Air Force's official account of its wartime camouflage achievements records that before 1938 very little was done to conceal airfields, although during the 1914–18 war decoy experiments had been carried out in both Britain and (more successfully) France.[2] Reorganisation of camouflage measures within the Forces was precipitated by the Munich Crisis, and the Air Force's mission was 'to be able to treat, in a practical and economical manner, distinctive features of an RAF station … so that at effective heights and distances in varying weather, in daylight and darkness, the pilot of an enemy aircraft would be deceived or confused'. The potential delay to a pilot's recognition of a target should either serve to eliminate attack or divert it.

In the early weeks of 1939 a special branch of the RAF Camouflage Section was set up in order to co-ordinate progress from conception to completion of design, and then to allocate the ensuing camouflage schemes to their respective Works Areas. It was decided that it would be necessary to enlist the expertise of 'an artist of repute' to direct and advise on works being carried out. The man selected was Norman Wilkinson, of dazzle ship fame, who was made an honorary air commodore in order to assist his liaison with the RAF. He remained Inspector of Camouflage for the Air Ministry until 1942 when his responsibilities were transferred to Colonel Turner's Department, which had been established in the autumn of 1939.

Colonel John Turner was a Royal Engineer veteran of the First World War who, on retirement from the Army in 1931, took up an appointment as director of works and buildings at the Air Ministry.[3] This work gave

72

Decoy aircraft: canvas stretched on wooden framework.
The National Archive

him unparalleled knowledge of the British programme of airfield development from the start of his job until the war, and for this reason he was asked to take control of the Ministry's programme of decoy and deception. His branch, labelled simply Colonel Turner's Department (never known by any more official title) and unlisted for reasons of secrecy, was based in London at Aerial House in the Strand. From there the Colonel began the task of deciding on suitable sites and equipment for a major decoy and deception programme.

At the time of Colonel Turner's appointment, the Air Ministry had already begun discussions about the potential of fake airfield sites, which could conceivably be situated on previously-planned but never-built satellites to permanent stations. The belief was that these, if constructed complete with tented 'accommodation' and dummy planes, would attract bombs away from the central base. The task of producing dummy aircraft was put out to tender to several large aircraft manufacturers, and although the resulting prices differed wildly, the lowest of them still amounted to twice the allocated budget, simply because the factories were geared up to manufacturing real planes and not lightweight imitations. As a result of this, the Air Ministry's Research and Development team conceived the inspired idea of inviting film companies to tender for these considerable and doubtless costly projects, and several studios took up the challenge. Both Alexander Korda's London Film Company and Warner Bros Studios submitted tenders but failed to gain contracts, and although Gaumont-British Studios produced a competent prototype, the best and most realistically priced proposal was offered by Sound City Films, based at Shepperton.

Colonel Turner's initial business was to recruit a team of officers, either newly-enlisted recruits or men from civilian backgrounds, and to arrange a training programme for them which was based at RAF No. 2 Balloon Department at Hook, Surrey.[4] Just after New Year 1940 the Target Plane Section was ready to accept the first dummy aircraft from Sound City, accompanied by the first wave of trainees who would eventually man the decoy aerodromes. The preliminary part of the training covered 'K sites' (day decoy airfields) and the second, 'Q sites' (night decoys, created by lighting, usually but not always supporting a day decoy). K sites were complete imitation airfields, strategically placed to lure the attacker away from the actual station. This was where the dummy aircraft, buildings and associated features came in, courtesy of the experts at Shepperton. The night decoys could be more randomly situated (therefore making the choice of location easier) as long as they were positioned on the estimated track of an enemy approach. They were usually about five miles from the real target. The initial 'Q' was adopted from the Navy's name for warships posing as merchant vessels ('Q ships'). Q sites used a system of electric lighting mimicking real airfield lighting, controlled from an independent dug-out shelter.[5]

The next mission for the Colonel's department was to perfect the creation of 'SF sites'. This initiative was triggered by the bomb devastation of Coventry. Codenamed 'starfish', SF sites (the initials standing for 'special fire') were designed to draw a second wave of enemy bombers away from a newly bombed target (often a town), under the impression that they were following up the target's first wave of attack. These decoys proved a huge success throughout the Luftwaffe's bombing campaigns. By the end of the Blitz in early summer 1941, 140 starfish had been put into operation, and the work was subsequently continued in North Africa and Italy. A page of notes taken by Ashley Havinden from a lecture delivered by Julian Trevelyan in 1943 at Farnham contains a small section referring to 'Col. Turner's Dept', in which Havinden writes 'Decoys and Fires … do really work.'[6]

In April 1940 the Camouflage (Policy and Organisation) Committee put out a recommendation that night decoy lighting should be provided to protect industrial targets; however, there was no mention of day decoys. Colonel Turner had already been instructed by the Air Ministry to begin developing schemes for six dummy replicas of major aircraft factories, including those of the de Havilland Aircraft Company and Boulton & Paul. It is evident that the service and civil ministries were not

73

74

at that stage liaising closely enough to arrange particular responsibilities in camouflage policy, even though
the Policy and Organisation Committee was set up to
create interdepartmental harmony. Turner was advised
to consult the Central Camouflage Committee which, in
his view, knew 'very little about camouflage and nothing about deception'! (Reasonably true at this stage of
the war – the RAF was certainly first in the Forces to
work with deception in the form of decoys and
dummies.) His dissatisfaction, and the fact that he was
at a fairly advanced stage with what was obviously a
crucial plan, resulted in a review of the recommendation:
Colonel Turner's work was made the responsibility of the
Air Ministry, and the development of K sites continued.

In 1941 Colonel Turner's Department was handed
responsibility for the static camouflage of all RAF
stations. His officers' training in static camouflage,
originally begun by pure guesswork in the absence of
any experience, took place at Farnham with visits to
learn from the Leamington team's experience of static
camouflage. Training was given in gunnery, to enhance
the realism of K sites, and in electrical skills to help with
Q lighting schemes (the Air Historical Branch records
that 'considerable drive was necessary to induce the
civilian crews … to try and learn something about their
equipment'!). Shepperton Studios were the base for
theory on decoy production, while the practicalities
were demonstrated on various working sites. Later in
the war, with the lighting decoys becoming increasingly
crucial, those who had not attended a course at
Farnham were also sent on camouflage courses at

*Opposite: Hugh Casson,
watercolour showing the
construction of a decoy
hangar to divert attention
from the distant airfield,
1941.
© Hugh Casson. Reproduced
with permission of Sir Hugh
Casson Ltd*

Tunbridge Wells or Norwich when places became available. Eventually two specialist schools were set up, one in Richmond Park for training on appropriate equipment, the other at Shepperton covering engine maintenance, map-reading and site reconnaissance, as well as the building of dummy aircraft and the concealment of real planes. Recruits also underwent a certain amount of battle instruction at selected army training schools.

A memorandum issued by the Air Ministry in 1940 remarks that although it accommodated a specific branch dedicated to aerodrome camouflage, the execution of the actual work fell to a remarkably large group of personnel under the 'Superintending Engineer of the Works Area concerned'.[7] Working with the engineer would be '1 technical officer (chemist), 1 assistant surveyor, 3 camouflage officers' and an architectural assistant for camouflage – 'trained architects of particular artistic bent who had the necessary 'flair' and capacity for colour and form appreciation'.[8] A 'small Design staff in the Directorate of Works' at the Air Ministry would prepare 'a scheme for each station … [taking] account of the original natural features and the final layout of the station, its tracks, principal approaches and dispersal'.[9] The design was then approved by the superintending engineer and discussed with the appropriate RAF station commander and his works executive. Any issues which could not be solved locally were referred to Honorary Air Commodore Norman Wilkinson, and later to Colonel Turner.

In addition to its responsibility for creating camouflage schemes, the staff of the Design Section (W8: Design and Model Experimental Work) at the Air Ministry also carried out the training of camouflage assistants for the Works Areas. During the 'peak period' for aerodrome camouflage in 1942 the Section employed 15 assistants working under an 'architectural and engineering assistant, grade 1'. Their task was to disguise the 'living' surfaces of airfield landing grounds; the government's static camouflage principles would continue to be applied to the buildings. A paper on 'Aerodrome Camouflage' submitted by Norman Wilkinson to the Camouflage Technical Sub-Committee in May 1942, requests a 'realistic attitude' to the hitherto wasteful use of manpower and materials in aerodrome camouflage, highlighting the difficulty of concealment during contracted runway works which naturally rendered the station very conspicuous. He suggests that only one of the Works Areas fulfilled the criteria for conditions in which an effective camouflage job could take place.

There were 15 Works Areas in 1940, increasing to 20 in 1942. Each Area was staffed by one camouflage officer, one assistant camouflage officer, two draughtsmen and one clerk. The approved work would then be executed by a team supervised by one camouflage inspector and '6 Foremen of Trades'.[10]

In summer 1940, when the London Blitz was at its worst, the young architect Hugh Casson was accepted into the Air Ministry's Camouflage Branch (via the help of an 'influential friend') and started training immediately, bravely commuting from Chelsea to the Holborn

A drawing of the interior of
Hugh Casson's office,
which accompanied his
story 'Christmas Eve at
the Site'.
The Architects' Journal,
23 December 1943

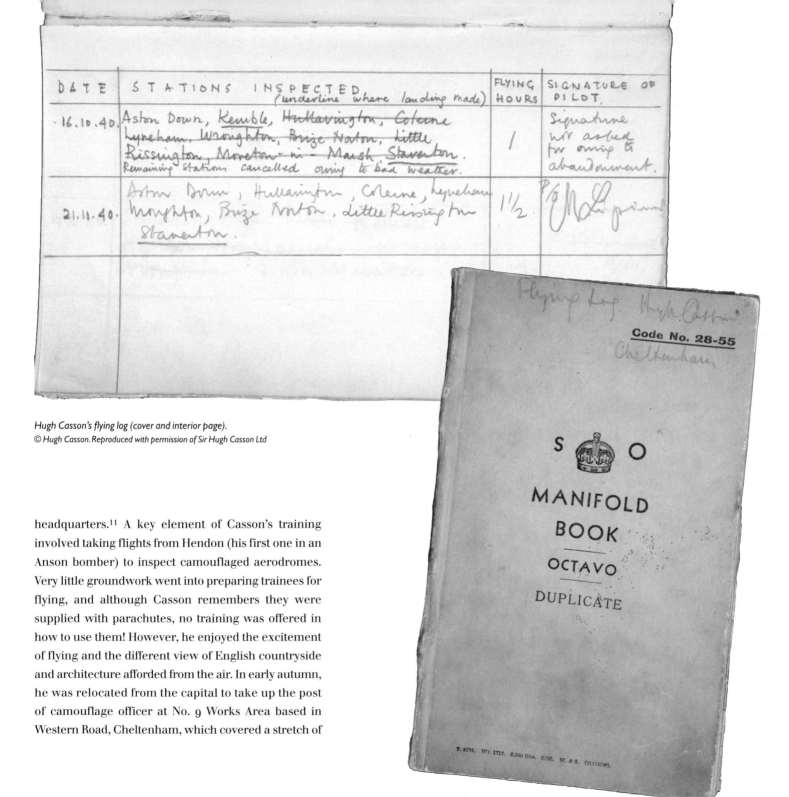

DATE	STATIONS INSPECTED (underline where landing made)	FLYING HOURS	SIGNATURE OF PILOT.
16.10.40.	Aston Down, Kemble, Hullavington, Colerne Lyneham, Wroughton, Brize Norton, Little Rissington, Moreton-in-Marsh Staverton. Remaining stations cancelled owing to bad weather.	1	Signature not asked for owing to abandonment.
21.11.40.	Aston Down, Hullavington, Colerne, Lyneham Wroughton, Brize Norton, Little Rissington Staverton.	1½	P/O EM Casson

Hugh Casson's flying log (cover and interior page).
© Hugh Casson. Reproduced with permission of Sir Hugh Casson Ltd

Flying Log. Hugh Casson.
Cheltenham.

Code No. 28-55

S ♔ O

MANIFOLD

BOOK

OCTAVO

DUPLICATE

T. 6774. Wt. 1715. 6,000 Bks. 5/36. W. & S. (2115410w).

headquarters.[11] A key element of Casson's training involved taking flights from Hendon (his first one in an Anson bomber) to inspect camouflaged aerodromes. Very little groundwork went into preparing trainees for flying, and although Casson remembers they were supplied with parachutes, no training was offered in how to use them! However, he enjoyed the excitement of flying and the different view of English countryside and architecture afforded from the air. In early autumn, he was relocated from the capital to take up the post of camouflage officer at No. 9 Works Area based in Western Road, Cheltenham, which covered a stretch of

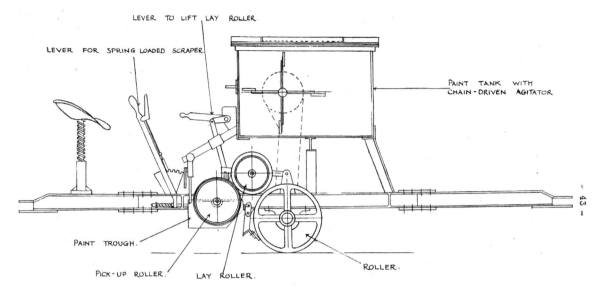

LEVER TO LIFT LAY ROLLER.

LEVER FOR SPRING LOADED SCRAPER.

PAINT TANK WITH
CHAIN-DRIVEN AGITATOR

Roller for applying paint to create 'hedgelines' and other landscape features.

PAINT TROUGH.

PICK-UP ROLLER.

LAY ROLLER.

ROLLER.

OVERALL LENGTH – 12 FT.

WIDTH OF TRACK. 7 FT. 7 INS.

PAINT ROLLER.
SIDE ELEVATION.

FIG D.

country roughly from Oxford to Cardiff and from Stratford-upon-Avon to Salisbury.

Casson was allocated a car for his work, which involved visiting airfields within his Works Area and ensuring that camouflage schemes were being carried out correctly (as opposed to Colonel Turner's work, which was to lure attackers away from these stations). His flying log shows the airfields he visited and revisited to assess work. Flying time amounted to around five hours per week, and flights were made in whatever plane happened to be in use that day – trips sometimes proved quite hair-raising, although Casson's

natural enthusiasm meant that he valued these new experiences. Flying hours were recorded and verified by the pilot's signature. The airfields at Brize Norton and Lyncham became frequent haunts, as did the smaller stations at Witney, Hullavington and Chipping Norton.

Most of the airfields, especially the newer ones, were constructed on farmland. It was therefore ironic that a proportion of camouflage work was carried out to make open spaces appear much as they had done before they were requisitioned. For example, hedgerows sometimes needed to be reproduced where hedges had already been removed to make way for a runway, or grounded planes were concealed by means of artificial trees, artfully created by suspending netting or painted canvas above them. Early landing strips were simply marked out on grassland, and the usual method of concealment was to paint them with coloured powders which imitated different agricultural surfaces. In 1939 the Paint Research Station at Teddington was commissioned by the Air Ministry to report on camouflage paints and some were found to be particularly effective. Companies such as Berger Paints were able to supply bituminous emulsions which successfully covered tarmac runways and felted roofs (in black these could be used for 'hedge' painting). Silicate paints adhered well to cement and concrete, while oil-bound distempers proved to be weatherproof and were therefore appropriate for walls and roofs.[12] Paints were continually tested and improved in terms of effectiveness and durability, and dusting powders for simulated fields, based on gypsum or chalk mixed with pigment and adhesive, were also standardised.

Hugh Casson also remembered spraying the grass 'with sulphuric acid ... to make a yellow field', and many other chemicals were brought into new use.[13] The need to arrest the growth of grass became apparent as painted 'hedgelines' disappeared with spring growth. Successful experiments were carried out with various chemicals, but the use of these was abandoned in 1940 when over-enthusiastic application began to cause serious damage to turf and vegetation.

Camouflage materials, particularly paint, were naturally in limited supply. The Air Ministry took on the complicated task of sourcing and arranging the supply and application of the necessarily enormous quantities of materials. Contractors were each allocated one or more Works Areas under the direction of the superintending engineer, who would supervise operations, managing plant and labour for each station.

Requirements for materials changed considerably with the introduction of hard-surface runways to accommodate fighter planes and heavy bomber traffic. Before the war, airfields were conspicuous because of their buildings; in this way they differed from civil airports whose runways were a giveaway. As a growing number of airfields were built with concrete and tarmac runways, it was found that from a reconnaissance plane at over 5,000 feet a runway camouflaged with paint was as easily identifiable as an unpainted one. It was necessary to find a successful way of reducing the 'shine' on the surface by means of texture. Experiments were carried out at guinea-pig airfields (at Stradishall with coloured slag chippings, and at Gosport with a similar treatment comprising chipped stone) and these were considered a success. It became evident, however, that the fighter planes' tyres were becoming excessively worn by these rather harsh surfaces, and various experiments to produce softer yet resilient coverings were carried out. Sawdust, woodchip, seaweed and pebbles were tried, but were found too soft and lacking in texture, although the first three made fairly good roof covering. Tan waste from leather tanneries was effective, but did not react well to wet weather.

Hugh Casson, watercolour of a yellow airfield, 1940. The field has probably been treated with sulphuric acid, which has also affected the grassed hangars. The small vehicle is possibly a paint-dispersing tank.

© Hugh Casson. Reproduced with permission of Sir Hugh Casson Ltd

*Eric Ravilious, Spitfires on a Camouflaged Runway, 1942. Before squadron unit markings were officially
stopped, landing strips were marked with paint. Watercolour on paper.*
RAF Museum

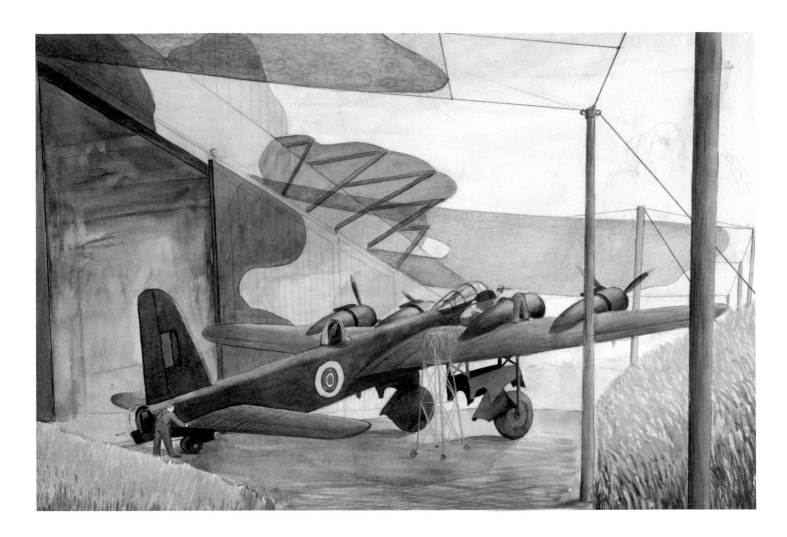

A treatment pioneered at Hullavington, one of Hugh Casson's responsibilities in Works Area No. 9, became the method adopted for fighter plane runways. This was a shine-absorbing 'open-textured bituminous surface' which became known as 'pervious asphalt'.[14] However, the fact that it was not completely waterproof made it only partly successful, and it was superseded by a material made of shredded rubber scrap. Eventually, after intensive trials, pulverised (hardened) wood chippings became the standard surface for fighter plane airfields, a material easy to source as wood was in plentiful supply. Colour was sprayed on with a dye manufactured for this purpose by ICI, or a special runway paint developed by Thornley & Knight.

B.V. Bishop, Artificial Trees, 1943. *Flat coloured surfaces create cover for an aircraft.* Imperial War Museum

Opposite: Hugh Casson, watercolour of a hangar disguised by both netting and organic grass treatment, (undated).
© *Hugh Casson. Reproduced with permission of Sir Hugh Casson Ltd*

Right: Drawings by Hugh Casson illustrating how to create 'natural' camouflage through planned cultivation, 1941.
© *Hugh Casson. Reproduced with permission of Sir Hugh Casson Ltd*

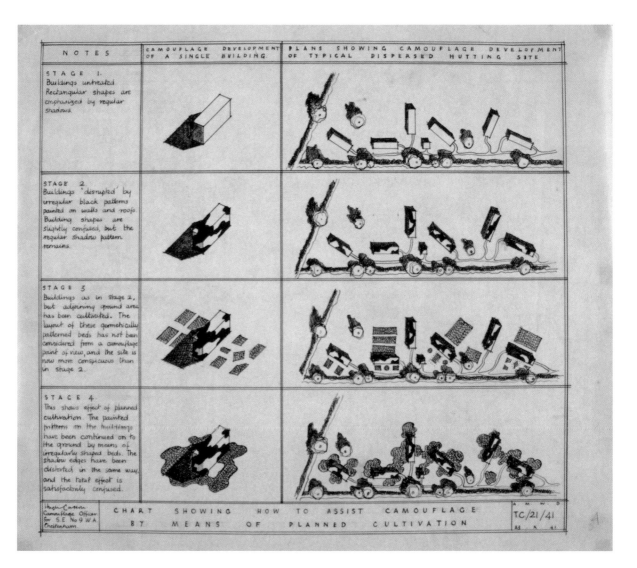

Notes recorded by Ashley Havinden at a refresher lecture by Julian Trevelyan in 1943, with illustrated detail.
Imperial War Museum

By 1941 it became clear that the war was not going to be over quickly, and that the manual methods of treating natural features such as grass surfaces and hedgerows were time-consuming and short-lived. Longer term schemes were therefore considered and put into action. Agricultural methods of seeding and mowing were used to create the appearance of fields, and because the illusion of hedges around the fields was of prime importance, hedge painting techniques were perfected. Ashley Havinden's notes from Julian Trevelyan's lecture of 1943 are succinct: 'Irregular edges of hedge therefore only possible by hand spraying. Knapsack sprayers: (vines etc …) Hand pumps: also can be improvised. Finally paint drag: (CD&TC *Bulletin*) both sides of grass therefore twice over ground. 30 cwt truck. Cattle trough.' The Air Historical Branch's summary of the RAF's wartime activities describes not a truck but a tractor with a drag net and spraying apparatus. New uses for all kinds of agricultural equipment were constantly being found.

Havinden's lecture notes contain a personal record of RAF camouflage projects throughout the war and describe the main methods used in England, as well as those developed for use abroad by the Royal Engineers. He looks at camouflage from the pilot's perspective, and makes comments aimed at helping to plan illusion and deception schemes: 'RAF hate: Pylons – Church spires – Telegraph poles – Trees – Hills' probably means that all these features make a target easily recognisable, but also indicate hazardous conditions for low-flying attackers.

The notes also record that in January 1943 Works

Cedric Kennedy, A Camouflaged Runway, 1942. Well-camouflaged sheds with a 'robin' (see page 88) posing as a bungalow. The airstrip has been disguised using both man-made and horticultural methods.
Imperial War Museum

88

Area camouflage officers were 'now taking in a lot of camouflage themselves'. This suggests that the workload of officers such as Hugh Casson had increased considerably. His drawings show meticulous attention to detail in the planning of many different disguises for airfield buildings, particularly for hangars ('robins') which might end up posing as garages, farm buildings or country cottages. Like Oliver Messel at Eastern Command, he must have delighted in being able to use his designer's eye for fantasy in creating these schemes! The detail went as far as to include features such as painted-on pillars, an earth closet, a washing line in use; a list of 'props required' includes '1 lean-to, 2 chimneys, 2 waterbutts, derelict farm machinery'. Materials were the same as those specified for civil camouflage; for hangars, hemp netting garnished with coloured jute scrim was often employed. Chicken wire with steel wool was laid on the ground to provide texture, but its use was limited as the cost was high.[15]

Casson was frequently required to produce colour sketches of schemes for contractors' use, and it was from this time that his skill as a watercolourist developed.[16] During his architectural training he had learned about applying colour washes, but for this particular job it was essential to reproduce plans containing exact detail and colour information. He found time to produce many exquisite watercolour sketches of work in progress and the effect of war on the countryside. In May 1944 he showed his work at a special exhibition entitled 'Gloucestershire in Wartime – an Architect's Impressions of Aerodromes and Architecture'. In fact,

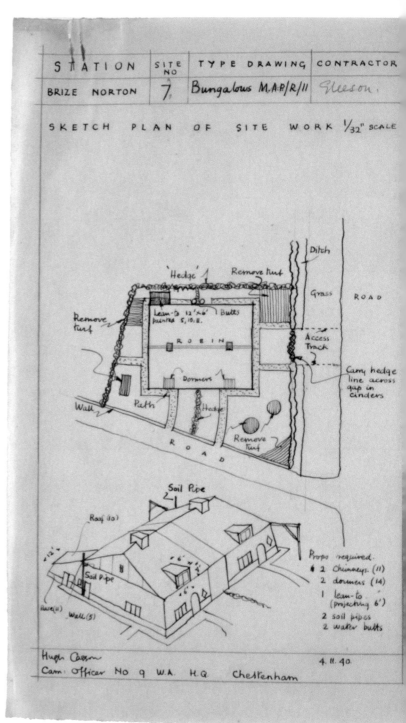

TATION	AREA	C.O.	TRENTHAM	CONTRACTOR
...IZE NORTON	12	S/L King	McTrusty	C & T PAINTERS LTD

TYPE	DRWG NO	PROPS REQUIRED	COMPLETION DATE	REMARKS
Barn	R/5			Substitute 12 for 10. 13 for 1.
Stacks	R/6			Reverse shadows.
Barn	R/4			Black for 10 + 1. (except bays)
Barn	R/3			Black for 10.+1.
Barn	R/3 (revised)	Site work. See special design.		Black for 10.&1.
School	R/8 (revised)	Site work + hedges. 1 stove pipe		Alter position of entrance door.
Bungalow	R/11	Site work + hedges. 2 chimneys. 2 dormers. 1 leanto. 2 soil pipes + butts.		
Barn	R/5			Substitute 12 for 10. 13 for 1.
Barn	R/4			
Chicken House	R/12			Roof to be 13 with black strips.

SUPER ROBINS

	TYPE	DRWG NO	PROPS REQUIRED		
2	Dutch Barn	S.R/1			
	Stone Barn	S.R/2			
	Stone Barn	S.R/2	Site work.		
	Dutch Barn	S.R/1			

...a Casson	CAMOUFLAGE 'S' SHEDS	DETAILS		30.9.40. 21.10.40.
...n: Officer	No 9 W.A.H.Q.	Cheltenham.		

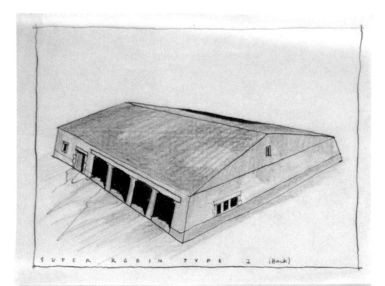

SUPER ROBIN TYPE 2 (Back)

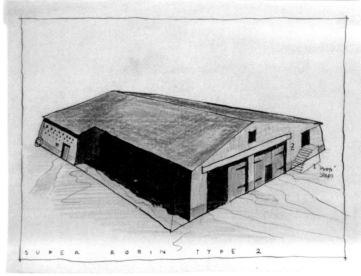

SUPER ROBIN TYPE 2

Left and opposite: Hugh Casson's 'robin' plan for disguise as bungalows.
© Hugh Casson. Reproduced with permission of Sir Hugh Casson Ltd

Above: 'Super robin' disguised as a barn with pillars.
© Hugh Casson. Reproduced with permission of Sir Hugh Casson Ltd

even while he was at Cheltenham, Casson managed to write and illustrate various magazine articles and also to act as a visiting lecturer at Cheltenham School of Art.

In early 1943 air camouflage policy was changed to concentrate mainly on potential attack to the east coast of Britain. Work at most of the designated Works Areas was relaxed or discontinued, so that only maintenance and remedial measures were undertaken. Along with many other camouflage officers, Hugh Casson's work came to an end, and he subsequently found a position at the Ministry of Town and Country Planning. In *The Architectural Review* (1944) he asks, 'when peace comes', would 'the aerodromes and camps ... strut once more across the countryside in the strident pinks and livid greys of brick and asbestos', or would they be 'carefully and inconspicuously sited, ... well built and not too untidy in their habits'? In other words, the valuable lessons learned from airfield camouflage could be taken forward to be beneficially used as a matter of course in the rebuilding of post-war life.

Opposite: Hugh Casson, watercolour of barrage balloons in a Cotswold village (undated). © Hugh Casson. Reproduced with permission of Sir Hugh Casson Ltd

91

CHAPTER SIX

Army Camouflage

FOR MANY PEOPLE, the words 'army camouflage' conjure up an image of sombre green and khaki patterns applied to fatigue clothing, tanks and tents, or men with painted faces and twig-garnished helmets crawling through the undergrowth. However, army camouflage in the Second World War probably took on the most complicated and widely adapted measures of the discipline in any of the Forces. Camouflage measures were required not only for disguising static installations and weaponry, but also for army defences on the move, encompassing the complete battle scenario from location and equipment to the concealment of the men themselves.

A secret War Office document circulated in October 1941 lists 'Responsibilities for Camouflage and Decoy Work'.[1] This divides the subject into two sections, 'field' and 'static' camouflage. The former was the term used to describe measures taken to conceal every element of an army encampment – the camouflage of headquarters, defence work, positions of personnel, vehicles and weapons – precautions which should automatically be taken by any army unit when setting up for combat. 'Static' army camouflage, as in the civilian world, covered concealment and disguise of installations including 'camps, hospitals, stores depots, ordnance workshops', and all military-related establishments which were potential enemy targets. Within these two groups there were differing levels of camouflage application – 'simple' camouflage, which described easily achievable methods such as the use of netting or paint techniques; and 'complete' camouflage for areas requiring a specially planned scheme designed by experts.

The War Office's section overseeing general policy for field camouflage was also given responsibility for training and development at Farnham. David Medd remembers that while there under the management of James Gardner he was appointed to section C43, the development wing or 'practical end' which carried out 'more serious camouflage development than the other unit populated by RCA people' who were making installations look quite peculiar by applying odd shapes and colours to them.[2] The development wing was responsible for studying tactical requirements and creating prototypes of all experimental devices for deception and concealment. AA and CD Directorate was essentially in

Eric Ravilious, Channel Searchlights (1), *1941. Watercolour on paper.* ©The Royal Pavilion, *Libraries & Museums, Brighton & Hove*

charge of the static camouflage (to include decoy work) of anything that could be a target from the air; and the engineer-in-chief's department had the remit of giving technical advice, costings and execution arrangements wherever needed. Furthermore, the Home Forces, AA (Anti-Aircraft) Command and Army Commands were responsible for the origination and maintenance of their own field and static camouflage.

In addition to the staff officers and advisors trained in field camouflage at Farnham, each Command also had a dedicated static camouflage officer. The CD&TC was responsible for instructing regimental camouflage officers and instructors from other schools, and it had a further obligation to produce military training publicity for all aspects of deception and decoy.

There appear to have been six main Command headquarters under General HQ Home Forces: Eastern, South-Eastern, Southern, Western, Northern and Scottish Commands. Some of these accommodated camouflage training schools, but their main work was to oversee camouflage arrangements in their respective areas, with special emphasis on coastal and fixed defences. In March 1941 Colonel Beddington proposed that each Command HQ would have one general staff officer and one staff lieutenant '(for fixed defences)'.[5] Other ranks were to consist of '2 batmen-drivers, 1 clerk RE (may be a Corporal), 2 draughtsmen RE Arch'. Vehicles provided were '2 cars, 2 seater 4 wheeled'.

As an example of specific staff arrangements, the staff list dated 29 March for HQ South Eastern

Edward Seago, letter from HQ 5 Corps to Brigadier Ritchie, and drawing of pillbox, September 1940.

The Problem

How does the air observer recognise a vehicle?

WHAT DOES HE SEE ?

Light

reflected from upturned surfaces.

Shade

on the large vertical planes.

Shadow

contained and cast by the form of the vehicle.

Contrast

These contrasting areas of tone can be made less obvious to the air observer by paint.

5

Application

THE BASIC PRINCIPLES laid down apply to the painting of every vehicle.

FOR CONVENIENCE, two colours only are generally used. For England and Northern Europe the light colour may be

Khaki Green No 3

or

Standard Cam Colour No 2

The dark paint should be

Standard Cam Colour No 1a

in some countries the problem is altered, though the basic principles remain the same. The shine from upturned surfaces will be no lighter than light desert country, therefore these surfaces will require darkening to a much lesser degree, or possibly not at all. Depending on the tone of the roof, the light toned pattern on the vertical sides must be made lighter. A very light paint will often be required.

REPETITION of one pattern on a collection of vehicles should be avoided. The following diagrams, which maintain the basic principles of vehicle painting are a guide, three types of pattern are suggested.

9

Army vehicle camouflage instruction pamphlet for training purposes.
Imperial War Museum

Command lists Major Pavitt as General Staff Officer, and two sub-lieutenants (each responsible for a particular coastal area). The Home Forces Corps headquarters answering to that Command were 12th Corps whose general staff officer was Captain John Lewis, responsible for the Kent area; 4th Corps, covering Surrey and Sussex, and Aldershot area HQ, which looked after any relevant general duties in the area. Corps were responsible for their own camouflage schemes, which were often limited to field work, and not complete camouflage coverage of their area. If the army corps were to be billeted elsewhere the GSO would probably stay and become area camouflage officer. Local anti-aircraft camouflage matters were the responsibility of the Command camouflage officer.

Subordinate to the corps camouflage officer was a unit camouflage officer (often trained on a Command or corps camouflage course). This officer's duties included the day-to-day management of the equipment and materials, the quality and maintenance of the unit's schemes and installations, and the organisation of in-house training for the troops. All these responsibilities are detailed in a 13-point list entitled 'Duties of a Unit Camouflage Officer'.

During the latter part of 1940 and early 1941, when Britain was under constant but as yet unfulfilled threat of land invasion, camouflage teams had time to experiment and develop methods of deceiving the enemy. Although official instructions would filter down from CD&TC through the Commands to the officers of the corps, and finally to the unit camouflage officers, much ingenuity was necessary to put ideas into practice. The supply of key materials was obviously of great importance, and it seems remarkable that sufficient resources

Roland Vivian Pitchforth, A. A. Battery, 1943, shows one of the many uses for netting. Pitchforth was a pre-war tutor at the RCA. Government Art Collection

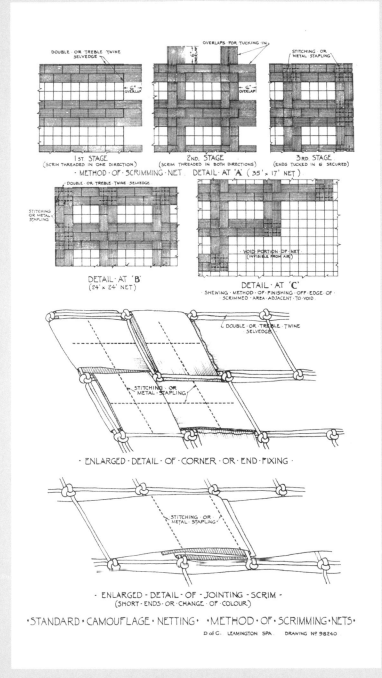

Diagram showing a method of scrimming camouflage nets.

were often readily available (again, by official order from factories with special wartime contracts). When suitable materials ran out or became obsolete, imagination came into play and sometimes the most unexpected methods proved successful.

Anti-aircraft guns were the most vital form of active weaponry in the early years of the war, when the sky was the main battlefield. These were generally disguised with 'scrimmed' netting – sheets of fishing-type netting 'garnished' with scrim, a roughly-woven hessian material. Resistance and fraying tests were carried out by stretching netting over a wind-tunnel to find the strongest specimen. These included a standard 'BG' type (apparently BG stood for 'bloody good!') as well as one that had been reinforced with 20 per cent sisal. Suppliers included such well-known textile manufacturers as Wilton Carpets, and furnishing specialists Waring & Gillow and Maples. Generally nets were around 66 square yards in area, and constructed from three-inch mesh garnished with three-inch scrim dyed green, red and light brown. Garnishing was achieved by interweaving straight strips of scrim (two rows scrimmed, one unscrimmed), or through the more time-consuming and less economical process of adding bow-tied knots. The necessary attributes of the resulting material included high density for weight, resistance to handling and weather, compactness, flexibility and convincing texture – colour had to be relatively dark and fade-resistant. Netting was obviously the most efficient way of disguising guns (as well as tanks and most other vehicles), although the writer

97

Evelyn Dunbar, Convalescent Nurses Making Camouflage Nets, *1941.*
Evelyn Dunbar was a student at the RCA from 1929 to 1933. As one of the few female
official war artists, she was commissioned to record women's activities during the conflict.
Imperial War Museum

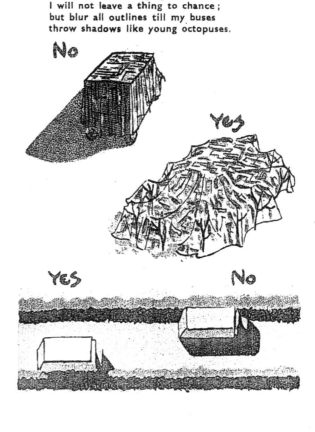

or post, or overhanging branch.
I will not leave a thing to chance;
but blur all outlines till my buses
throw shadows like young octopuses.

No

Yes

YES   No

Said Thomas: "The idea's immense!
Magic my foot! It's common sense.
I'll spit and polish day and night
in barracks, but when I've to fight
I think I'll dull my buttons down!
My boots shall be a muddy brown.
If British flesh shows pink, I must
black-out my handsome phiz with dust,
(and rub in more when I perspire).
I'll trim my hat with chicken wire
and twigs and leaves to break the line.
No longer shall my badges shine;
for though confusing it is right
to change my habits when I fight.
When not in action spit and polish!
When out for blood all brass abolish!
So, letting all equipment tarnish,
I'll spread my nets and thread my garnish,
(thick-centred, thinning to the edge),
and always drape it from a hedge

Page Eight

Page Nine

99

*Clemence Dane's training poem, 'Hide and Seek', commissioned by CD&TC. The
illustrations are by Fougasse, the pseudonym of ex-Royal Engineer cartoonist Cyril Kenneth
Bird, known for his wartime public information posters. He later became editor of* Punch.
Imperial War Museum

Norman Scarfe recalls that camouflaged guns, not surprisingly, were extremely difficult to operate.[4] Three-inch mesh netting with two-inch wide strips of scrim, sometimes interspersed with bows to increase texture, were specified for artillery use. Oily rags and coir could be introduced to offer surface variation.

Towards the end of 1941 netting was considered such a crucial element in camouflage development that several major textile factories, including Courtaulds, were under constant commission to produce vast quantities of the material. In July 1942 the Ministry of Supply wrote an angry letter to Wing

100

Subject:- "Cullacorts" Foliaged Wire Screening.
Ref:- CE/8/3/1/65.

A discovery has been made with the use of
"Cullacorts" material cut and applied in a special way which
gives a completely convincing representation of living foliage,
and it is extremely valuable for disguising pillboxes where it
is necessary to make them merge into hedges, trees, bushes etc.

Hitherto nothing has succeeded in giving
satisfactory results in this way, steel wool giving entirely
different effect and only useful as representing grass mounds
etc. Naturally, therefore, it is not successful camouflage
to use steel wool for continuing hedges and trees.

As there exist a large number of cases where it
is expedient that this new imitation of foliage should be used,
and the work is being held up until the "Cullacorts" arrives, it
is of great urgency that it is made possible for the material
to be obtained in order to avoid considerable delay and added
expense.

An illustrated description of this process
and photograph with interpretation are attached.

(SD) Oliver Messel -

H.Q., 3 Corps,   Lieut. R.E.,
Home Forces.   for Corps Camouflage Officer.
4 Nov 40.
OMS/EM.

*A letter and sketch by Oliver Messel, promoting the
use of 'Cullacorts' material.*

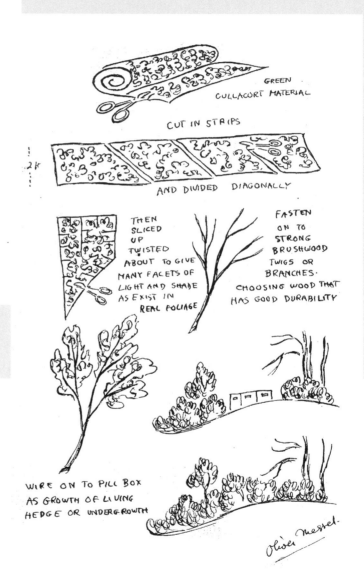

Commander Cave-Brown-Cave, complaining that although the Camouflage Technical Sub-Committee had nine months previously approved the importance of netting as a main component of the official camouflage recommendations, there was some surprise that the Air Ministry and the Army were now indicating that BG netting was not a high priority. This represented a complete change of policy. The factories were now ready to produce the requested 14 million square yards, having halted all other production.

As a result of the letter, a representative from the Ministry paid a week-long visit to Farnham to study the Army's needs, which had altered because of the changes in camouflage objectives over this period of time. It was decided that nets could continue to be used for the following: artillery and vehicle covers, with leaves and bracken added into the scrim (shrimp net was recommended for tanks as the mesh was smaller in size and less 'catching'); static overhead canopies, and mounded covers for machine-gun posts – although steel wool was better for this purpose; and screens to hide troop movement and gunflash. Large nets garnished with painted hessian or coir were used for ground cover. The Ministry representative was impressed by his visit and wrote a letter recording his grateful acknowledgements to Colonel Buckley and staff.

Although netting was a fairly simple answer to the problem of disguising large objects of irregular shape, its use was often less than satisfactory. In *The Fortnightly DO* (issue no. 10) there is a reference to a 'tacit belief in the covering power of an ungarnished net'.[5] It rather insultingly suggests that drivers and other untrained troops thought that throwing an untreated net over a vehicle would magically make it disappear, thus providing safe cover. Geoffrey Barkas relates in *The Camouflage Story* how he discovered this belief amongst troops he was training in Northern Ireland, and as part of the training package he composed the wonderful 'Sad Story of George Nathaniel Glover'. The popularity of this poem evidently did much to help save army vehicles and military lives.

It was eventually found that in many situations BG nets were not sufficiently durable and a substitute was necessary. The 'Cullacorts' material used so extensively in civil camouflage came into its own, particularly as a method of creating dummy trees and hedges. Its structure of wire netting garnished with feathers and other adornments, often layered with BG net, was tougher and could retain shape independently, and as a bonus, the manufacturers were willing to supply it ready-painted and undertake its installation if needed. Oliver Messel was a great protagonist of 'Cullacorts', especially (of course) where it could be used to disguise pillboxes. A letter he wrote in late 1940 on behalf of HQ 8 Corps, Home Forces, urges the increased production of the material, particularly because steel wool which was sometimes used as a substitute gave an 'entirely different effect'.

Steel wool was successfully used to cover larger installations, as was demonstrated by its extensive application in civil camouflage, and was appropriated by the Army for disguising small defence posts and some artillery. However, most gun batteries needed to be immediately to hand, which meant that it was not possible to develop three-dimensional disguises and covers for them (except when in storage). Paint, plainly, was the answer for camouflaging weapons which might be needed without warning, and it must not be forgotten that paint was the basis for most British camouflage early in the war. For example, in a letter dated April 1941 to Major Edward Seago at HQ

102

Headquarters,
8th Corps,
Home Forces.
3rd. April. 1941.

2/8888.
212

A:- CE/8/C/28/9.

Dear Major Seago

      At the request of Lt-Col Beddington O.C., 24th Fd. Regt. R.A.
I painted two 25 pdr. guns in two different schemes, one rather similar
to the pointillist model truck that you showed us, the other in a
disruptive pattern using countershading.

      The O.C. is anxious to paint the other guns of the Battery
in a similar way, and I promised that I would try and procure the paint.

      Could you authorize the purchase of the following quantities
of matt gritty oil paint.

        32 lbs.........Light green.
        32 lbs.........Dark green.
        32 lbs.........Light brown.
        32 lbs.........Dark brown.
        16 lbs.........White.
        16 lbs.........Black.

      It would be most satisfactory if you could authorize the
purchase of this paint at Messrs Serle, 51, East Street, Taunton where
I have always bought the paint for my experiments and could supervise
the mixing of the correct shades.

      Yours sincerely

        J. J. Trevelyan S/Lieut (Camouflage)

Major E. Seago R.E.,
Command Camouflage Officer,
Headquarters,
Southern Command,
SALISBURY.
JOT/EK.
----

*Letter from Julian
Trevelyan to Edward
Seago, requesting
authorisation for the
purchase of paint.*

Southern Command, Julian Trevelyan at HQ 8 Corps requests authorisation to purchase six enormous tins of 'matt gritty oil paint' in exactly what one would expect as camouflage colours – two greens, two browns, white and black. He relates that he has tried out two camouflage options on 25-pounder guns – one pointillist in technique and the other disruptive. Seago's rather stiff reply suggests that he try official suppliers rather than the paint merchants he has requested, but does indicate an interest in the results.

In February 1942 the unit camouflage officers were renamed instructors, so that it was clear that they were responsible for the camouflage measures to be carried out by their fellow officers and troops. Not least in their duties was the task of concealing the men themselves, even though the threat of field combat seemed at that time quite remote. *The Fortnightly DO* of that month reports on the importance of work done under cover of darkness, although 'the cover that darkness affords is actually very limited'.[6] A list of notes discourages the 'wearing of light-coloured equipment', and suggests darkening the face and hands. The subject of methods and materials which could make skin colour less conspicuous had been under review for some time by CD&TC at Farnham, although up until this point they had not been considered of great importance. However, the subject was obviously of interest, and Colonel Buckley made a summary of the tests carried out on various possible materials. These included soot, cocoa and cow-dung; unsurprisingly, reports of 'dermatitis and skin affections' soon materialised!

One stands against a dark background (bushes or trees). He shows as a light object on dark. The other stands against a light background (grass, plough), he shows as a dark object on light.

Why is a man seen ? Because he is darker or lighter than his background

(b) The men change places to show that the fact that they are dark or light depends not on themselves but the BACKGROUND.

(c) The men move across the front. A man shows to a greater or lesser extent according to what cover is behind him.

2. Why can these men be recognised as men ?

(i) By the characteristic HEAD and SHOULDER shape. A target in the distance is pointed out which has only the rough shape of a man's head and shoulders. It immediately suggests a man, because that head and shoulder shape has been recorded in our minds as characteristic of MAN since we first learnt to see.

One of the two men moves against a bush or into grass and takes up a position which distorts his head and shoulder shape. He is not recognised or seen so easily.

The two men then lie down. The head and shoulder effect remains, though less obvious. Even a bump on the ground of a certain size is to all intents a man. If there is sunlight, and suitable ground, one man can be shown facing into the sun, and the other away from it, and the difference between the two discussed.

(ii) By the spidery effect of limbs (the space between the two legs and between the arms and body).

(iii) Because some small part of a man or a man's equipment which is too dark or too bright catches the eye and is recognised.

A man is shown standing in full equipment close to the class. They discuss the various points which may catch the eye (Inf. Trg. 44, Part

6

VIII, p. 10). The class is then told to pick out various men on the ground who are invisible except for one thing which is showing, and to say why they recognise them

A man has been placed 100 yards off in front of a wall or hedge where he merges completely with his background. He works the bolt of his rifle and the class get the direction by ear, but can pin-point him only by his slight movement.

The class may also be told to mark the spot as they would a cricket ball which has disappeared over the boundary. They are then turned about for a second, and afterwards must point out exactly where the man is.

3. Men are shown standing in the open at various distances from the class (100 yards, 200 yards, 300 yards, 400 yards). The class judge the distance of each man, and are questioned as to the amount of detail which they can see at each distance. The men move across to a different background (bushes, hedges, undergrowth) and the process is repeated. They then move to backgrounds where they are hardest to see, and the class examines each man again. The class may be turned about during each move, and the distances which have to be judged varied. (This exercise should be repeated on different days and in different lights. On

(A) FACE

(B) HANDS

(C) HELMET

(D) EQUIPMENT (Webbing, Gaiters, etc.)

(E) MOVEMENT

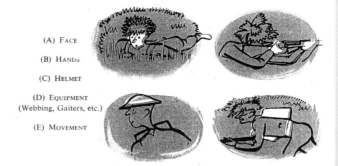

subsequent occasions it can be combined with other lessons, for example, as an observation exercise in which different types of equipment and signs of military activity are laid out—spoil, wire, " ammo " boxes, a carrier, a truck, real or dummy guns, tanks, etc.).

DEMONSTRATION II   USE OF BACKGROUND

The class now has a thorough understanding of why individual men show, and why they are able to recognise them, but even during

7

103

'The Enemy Watches', an excerpt from a personal concealment training pamphlet, illustrated by Charles Mozley.
Reproduced with permission of the estate of Charles Mozley

Eventually, a number of major cosmetic companies took up the challenge of manufacturing products suitable for use on the skin. Cyclax had already developed a highly popular cream for women which closely simulated the texture of silk stockings, an elusive wartime luxury. The company went on to create a dark-brown version of this which seemed to fulfil all the appropriate criteria for use in the field, and this was sent to various camouflage centres and schools, all of whom reported a favourable reaction to it. Colonel Beddington endorsed its success and wrote to the War Office encouraging its manufacture and distribution.

104

Throughout 1942 Colonel Buckley delivered 're-fresher' lectures to army camouflage officers and instructors at Farnham. It must have been frustrating for men who had been researching camouflage since the start of war to find that their plans and developments often seemed to go nowhere, although civil defences and airfield camouflage had with some success been protecting vulnerable major installations. The real action was taking place abroad, where the men and women of Special Operations were adapting everything they had learned to a different and very testing scenario. Ashley Havinden's notes at Buckley's lectures somehow convey a distinct lack of morale. Excerpts from a talk in February 1942 are as follows: 'Well over a year with little or no question of attack … Op. cam. losing maintenance interest'.[7] Buckley's emphasis was on the changing direction taken by combat developments, and the question of training in the right kind of camouflage. Evidently, the 'high ranks' at the War Office believed that camouflage was of little use, and it would be necessary to give them a very clear picture if they were to be convinced. The new direction should probably be towards misconception rather than concealment (which was everybody's everyday responsibility). The key words of the moment were 'dummy', and 'misdirection', as already potently illustrated by civil and airfield decoys developed by the camouflage team.

Colonel Buckley obviously wanted to boost the confidence and morale of his officers, and the aim of the refresher courses was to generate optimism. But in his lecture on 22 January 1943 the tone was at a distinctly low ebb, as again recorded by Havinden.[8] Random excerpts from his notes record - 'Posn [position] at present: not so hot'; 'No bullets or shells; Cam unimportant'; '1 year ago, Cam had support everywhere plus materials'; and 'Going downhill: further yet'. The gist of this talk was that the work which had been carried out by the camouflage team so far had concentrated mainly on concealment; the possible invasion of Britain in 1940 had made 'direct and simple calls on Camouflage officers'. However, the CDCE had already made big staff cuts, and inspectors were now considered 'redundant'; all tactical plans must now concentrate on the likely invasion of Europe.

Havinden's notes continue – 'Supposing invasion of Europe: … Big job – Hell of a lot we can do if given time.' Buckley's lecture emphasised the importance of altering the camouflage designer's frame of mind to deal with new problems. Display (in other words decoy) was the new direction, rather than simple concealment which had so far been the main method: 'setting of stage is really [the] important part'. The conclusion to this lecture was that the lessons learned from the victory at El Alamein would shape the thinking behind all new camouflage schemes, which must be aimed at *misleading* the enemy rather than attempting to conceal major targets.

A letter from Victor Stiebel (at Farnham) to all Command officers, written on 15 July the same year, confirms that this concept was immediately put into practice.[9] The policy was that 'now that the development of Devices for Visual Misdirection is getting into its stride', the work in progress at the CD&TC should be co-ordinated with that of the relevant Camouflage Commands and corps. Prototype models 'of any item of Army Equipment that the War Office might wish dummified' would be developed at Farnham and, once tested and approved, manufactured in bulk. In addition, it would be important to develop 'the use of decoys and their associated features in the field'. These would be for the use of trained camouflage units only; units requiring 'enemy fire-drawing decoy' should improvise with locally sourced materials to a broad plan outlined by the CD&TC.

The circulation list for this instruction contained not only Havinden's name but also Camouflage Command officers, including captains Gabriel White (later at the Arts Council) at Eastern Command, Oliver Messel, Basil Spence, and William Maving Gardner – sure proof that some interesting design ideas would materialise!

As noted by Ashley Havinden, 'Misinformation of enemy's intelligence is last touch': the display of the false was the new direction. The camouflage team was back in business.

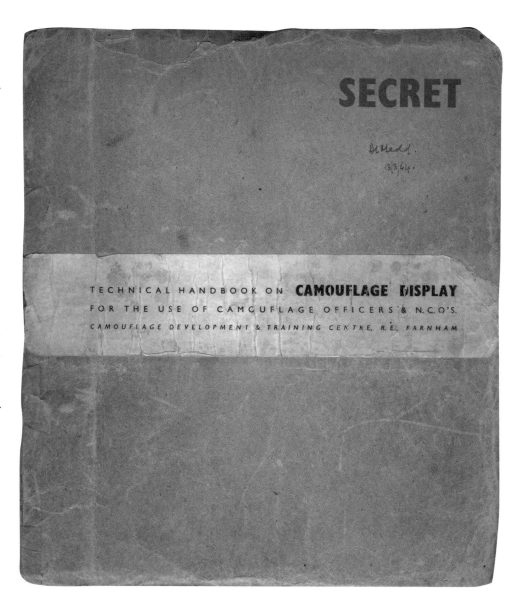

Front cover of the Technical Handbook on Camouflage Display, originally conceived by James Gardner as the 'Pneumatic Man's Bible – a catalogue of dummies and how to deploy them.' Evidently the War Office refused to print this.
Property of David Medd

CHAPTER SEVEN

Desert Camouflage

AN ACCOUNT OF BRITISH camouflage activities in the Second World War cannot be complete without extensive reference to Special Operations, and in this case the activities of the Army in North Africa. The operations carried out to foil first the Italian army and then the Germans during the years 1941 and 1942 were to prove the success of much of the camouflage theory and practice initiated at Farnham in the early stages of the war, and were also to pioneer a new point of view for war in Europe, that of the arts of decoy and hoax.

The war in North Africa began in September 1940, a 'low point' in activities at home when, according to Stephen Sykes, there were 'no true operational areas other than the Western Desert'.[1] Although training at Farnham was originally aimed towards possible combat situations on British soil, war developments were such that Royal Engineers trained at the CD&TC in British camouflage were soon billeted to Egypt to use their knowledge and resourcefulness in a totally different physical environment.

Marshal Graziani's Italian forces had landed in Tripoli and advanced along the coast, occupying all major settlements and eventually invading Egypt.

General Sir Archibald Wavell's much smaller British force, seriously outnumbered in both troops and artillery, finally dug itself in at Sidi Barrani to await reinforcements en route from England. To convince the enemy that they had a superior army of troops and tanks, British troops created a battery of dummy tanks and artillery built with any scrap materials on which they could lay their hands. These vehicles and guns were rendered more realistic by means of noisy flashes of fire emanating from a length of metal piping attached to each dummy gun. This apparently convinced the Italians that they were in the presence of a dangerous force, and in the words of Jack Miller, a North Africa veteran who returned to Farnham in 1943 to lecture on deception in the desert, Wavell's idea 'set the stage for camouflage [in the Middle East and for the rest of the war]'.

On New Year's Eve 1940 a convoy of Royal Navy destroyers and Royal Engineer troopships arrived at Port Said to report at GHQ Middle East Command, Cairo. The party which would take charge of camouflage operations was headed up by cinematographer Colonel Geoffrey Barkas, and artists John Hutton, Peter

Opposite: Little-Known Units of the Western Desert – Extraordinary Sartorial Field Company. *An insight into the unorthodox army field dress worn by members of the British camouflage team in the Middle East.* Brian Robb, from My Middle East Campaigns, Collins, 1944, reproduced with permission of Quentin Blake

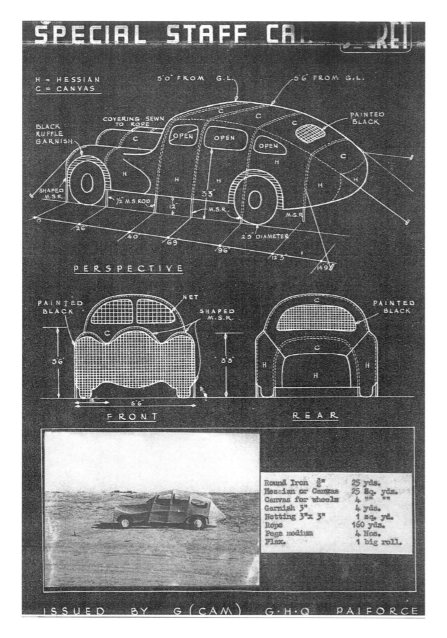

Plan and photograph of a dummy staff car.

108

Phillips and Blair Hughes-Stanton were the other three trained camouflage officers. Their mission was to research and develop camouflage schemes which would protect forces in desert warfare, a scenario which could hardly be more different from the British terrain, first at Shorncliffe, Kent and then at the Royal Artillery camp at Larkhill, Salisbury Plain, where they had received their training. However, the principles would remain the same: to conceal or disguise telltale marks made by an army in action.

Concealment in the desert posed a new set of problems for the camouflage team. Unlike the English landscape, there was no natural cover in the form of trees, brush or undergrowth; illumination issues were entirely different too. However, on a maiden reconnaissance flight from Cairo to the west, Barkas records that the desert's regions contained various natural patterns which would prove useful to the camoufleur.[2] A new descriptive terminology began to develop: land configurations were given names such as 'Polka Dot' and 'Figured Velvet'. The desert was not simply sand; there were stony and rocky areas too, and Julian Trevelyan's lecture back at Farnham records 'flowers carpeting the ground going westwards towards Tobruk'.[3]

In the words of Julian Trevelyan 'concealment from the air was almost an impossibility', and for this reason army activities would need to be disguised in new ways.[4] General Wavell's false artillery was the route forward, and the camoufleurs' inventiveness rose to new levels. However, it was found that many of the

basic methods explored in England could be adopted in the desert. Stephen Sykes writes that even as Colonel Barkas was adapting the guidelines contained within his booklet *Concealment in the Field* for use on foreign terrain, desert army units were arriving from England, dutifully but pointlessly covering their vehicles with the green and brown nets that had arrived with them, and rendering the encampments highly visible. John Morton, who was already in Egypt with the Searchlight Regiment, got a transfer to the Royal Engineers and

109

Above: The camouflage team aboard Andes. *Shown here are John Hutton, Geoffrey Barkas, Peter Phillips and Blair Hughes-Stanton.*
Reproduced with permission of the estate of John Hutton

Right: 'Textile' pattern in the desert, as recorded in the No. 1 Middle East Training Pamphlet.

14

(b) **Velvet with some Pattern.**

Desert again, but this time with some sandhills and a little broken ground. The unevenness and the tendency of the hills to cast shadows are already creating a rather stronger pattern, as we might expect, for dark lines and areas on a light background are, in effect, a *pattern*.

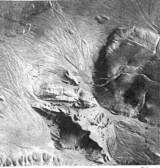

FIG. 13.

FIG. 14.

15

(d) **Wadi Pattern.**

This is how typical wadi desert country looks — the airman's view of the sort of country found near Bardia or Tobruk.

FIG. 15

FIG. 16

(c) **Another Desert Pattern.**

Notice the streak of dots caused by scattered clumps of camel grass and scrub. A few more spots of the same shape or size would not be conspicuous among or beside these.

(e) **Polka Dot Pattern.**

This is how the scattered thorn tree bush, as found in many parts of Sudan, looks to the airman. Any country where you have scattered trees on a fairly level or rolling plain will look something like this.

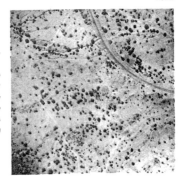

110

Little-Known Units of the Western Desert – Signboard Painting and Maintenance Company. *A humorous illustration of organised confusion. Brian Robb, from* My Middle East Campaigns, *Collins, 1944, reproduced with permission of Quentin Blake*

was later made a sergeant in a camouflage unit for the fictitious '9th Army' in preparation for possible defeat by Rommel's forces. He also remembers that rolls of green, brown, black and white hessian arrived for garnish, but the camouflage team managed to exchange these for 'sand-coloured' fabric. (Morton recalls that fabrics would mysteriously disappear only to re-emerge later on children wearing garments spirally constructed from strips of two-inch hessian!)[5] Another way of disguising the green hessian, remembered by William Murray Dixon, was to mix it with white garnish, and yet another treatment was to bleach it with sea water and slaked lime.[6]

Julian Trevelyan was sent out 'with a list of ques-

tions from Farnham' to find out how Barkas's team operated.[7] He recalls that the camouflage exercises carried out in Helwan's sandy environment were initially not always appropriate to the Western Desert, and that nets quite often had to be regarnished in order to adapt them for a different terrain.[8] Eventually, however, as the fighting developed, it became clear that nets were not enough, and as Dixon comments, 'All this type of work became obsolete.'

William Murray Dixon was a Royal College of Art scholarship graduate already stationed in Egypt. He was a sergeant in the Rifle Brigade (to which he was recruited 'on account of poor handwriting' – he had intended to apply for the Royal Engineers), when he was transferred to camouflage duties after Captain Peter Proud had enquired of the War Office whether any ex-RCA students were available locally in Tobruk. As a staff camouflage sergeant, Murray Dixon gave regular training lectures to Indian troops posted in the desert. His artistic abilities were put to good use in these lectures, during which he communicated with sketches rather than words as he had no knowledge of Urdu. His lectures on 'Camouflage: How the Enemy Sees Us' were well-received – 'the troops were very interested'.[9]

It became obvious as the fighting progressed that, because of the difficulty in hiding army operations, the most effective method of camouflage was that of deception. The advice to army encampments was to spread the installations and equipment about as much as possible, therefore eliminating the effect of a unit, and then to disguise each element. Where appropriate, the whole area could be made to resemble a settlement or small town (one such, developed as a scale model at Helwan, was unofficially given the very English name of 'Great Hoaxingham'!).[10] Dug-out trenches for the concealment of troops and weapons were easily disguised with sand, which could be mounded up to blur the outlines of any excavations. These were often further obliterated with spoil-covered corrugated metal roofs. As sand was in such abundance, it was natural that the camouflage team should decide to use it to its fullest extent, experimenting first in the area of disguise and then of deception. Glinting windscreens could be filmed over with oil, then sprinkled with sand to cut the glare. An imaginative use for the endless supply of sand was recorded in the January 1942 edition of *The Fortnightly DO*: when 10 tons of condemned Italian flour were found in an abandoned settlement they were mixed into a paste, applied to all the British tanks in Tobruk and then covered with sand. Apparently this treatment only took 10 minutes per tank. On a similar note, 2,000 gallons of Worcester sauce judged unfit for consumption by the troops were rescued by the ever-vigilant camouflage team and turned into a very effective paint thinner! Fake rocks were constructed from petrol cans and cement; another use for the cans being to crush them and crown them with a tin hat to make a dummy soldier's head. The ingenuity was unstoppable.

In all cases it was of course essential to be aware that tracks and spoil were probably even more notice-

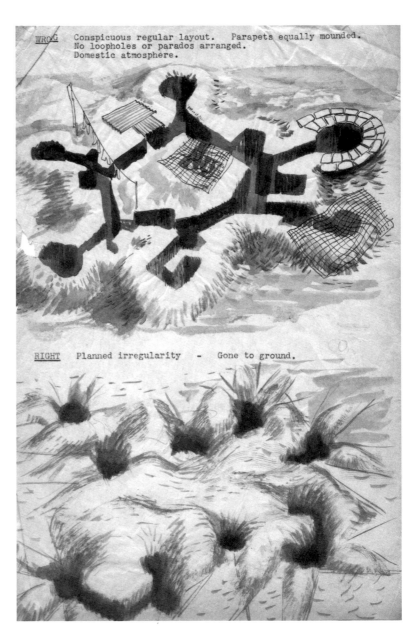

WROG Conspicuous regular layout. Parapets equally mounded.
No loopholes or parados arranged.
Domestic atmosphere.

RIGHT Planned irregularity – Gone to ground.

William Murray Dixon, diagram showing how to conceal a desert trench.
Property of W. M. Dixon, ARCA

111

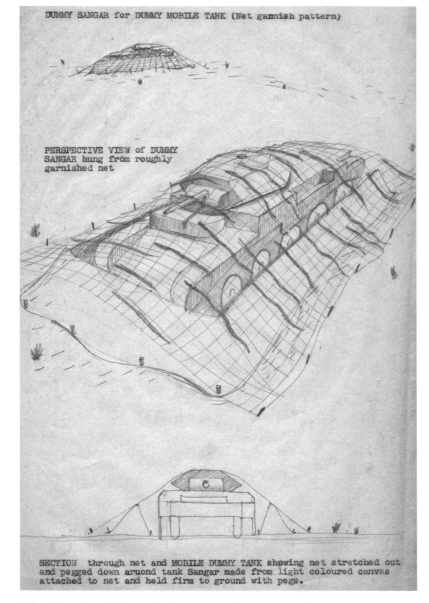

DUMMY SANGAR for DUMMY MOBILE TANK (Net garnish pattern)

PERSPECTIVE VIEW of DUMMY SANGAR hung from roughly garnished net

SECTION through net and MOBILE DUMMY TANK showing net stretched out and pegged down aruond tank Sangar made from light coloured canvas attached to net and held firm to ground with pegs.

112

William Murray Dixon's drawing of a dummy tank, itself draped in camouflage netting, is an example of the kind of double deception accomplished by the camouflage team. Property of W.M. Dixon, ARCA

able in the desert than on British soil. In addition, a further area where new rules had to be learned was the inevitable existence of shadow, which was a giveaway in almost any desert situation observed from the air.

The visibility of shadow played a crucial part in the work done by the painter John Hutton, who had been transferred to Egypt as a second lieutenant (in Julian Trevelyan's words, 'one of the most sympathetic camoufleurs ... out here ... desperately homesick, and yet supremely successful').[11] Hutton was put in charge of camouflage for the whole of the area surrounding the Nile Delta. His task was to disguise any new installation or defence amongst the dozens which were constantly materialising. New airfields were being constructed all the time, and as some of the most vulnerable sites their camouflage was crucial. Hutton's expertise lay in mural painting, and his answer to this dilemma was to create what he named 'shadow-houses' on the runways – *trompe-l'oeil* buildings appearing to extend the adjacent built-up area. These paintings were designed to confuse an enemy bomber pilot. Entire streets fronted with two-dimensional houses were cunningly created by means of paint, netting and tar-sprayed clinker along the runways and over roofs of airfield buildings, complete with their corresponding shadows. Although the shadows were probably only in the correct position for a short time each day, the general effect was so successful in confusing a low-flying pilot that the method eventually had to be abandoned – Allied pilots were themselves finding it difficult to locate the landing-strip and touch down safely.

A lesson learned from these shadow-houses was that desert camouflage and deception work needed much more maintenance than elsewhere. The frequent sandstorms would often obliterate the efforts made by camouflage teams, in an instant covering the painted surfaces with a thick layer of dust.

Through the use of dummies and decoys, false trails could be laid to trick the enemy away from awareness of troop movement. General Wavell disguised his tanks as lorries, and was able to strike when the Italians were least expecting it. Trevelyan remembers in *The Architectural Review* that throughout the desert campaign trucks were disguised as tanks, and tanks and guns as trucks. Dummy artillery and troops were created from any debris that could be found: sticks, canvas, petrol-cans, anything else that was left after combat. Owing to the difficulties of moving battle salvage (such as tanks or planes) the desert was littered with the debris of war, and gradually this too came to play a part in deceiving the enemy.

Peter Proud had arrived in Cairo in March 1941. Amongst the team of 12 camouflage officers drafted with him were Stephen Sykes, the magician Jasper Maskelyne (surely the perfect man for decoy and deception!), as well as John Codner and Robert Medley, both painters.[12] Around this time Rommel's formidable German army, which had landed in Tripoli in February, had advanced along the coast and overcome Agheila. Tobruk was under siege, with the British Army and the 9th Australian Division doing all they could to protect it. Proud quickly assembled a camouflage force and

tapped all resources in order to protect the city, whose only entrance for essential supplies was the harbour.

The camouflage team quickly bagged the harbour and its bombed and grounded ships, and used them to enact a kind of double deception. William Murray Dixon, working closely with Peter Proud in bombed-out Italian sheds with the detritus left outside, drew up designs to deceive the enemy. The destruction of genuine battle was extended to form dummy wreckage whereby naval supply boats ('A' lighters), arriving under cover of night from Alexandria, could be concealed under a real but artistically-enhanced jumble of scrap metal and canvas, further extended with the use of garnished netting simulating the beachhead. A similar mixture of methods was used to protect the water distillery which, built by the Italians, was obviously known to the Germans and therefore a sure target – after a bombing raid which fortunately did little damage, the camoufleurs were instantly able to 'wreck' it by digging dummy craters, scattering rubble and generally giving it a 'bombed-out' look with the use of paint, texture and smoke. A direct hit was proudly announced by the enemy.

During the siege of Tobruk the Germans held almost complete air superiority, while Britain's air power was limited to three Hurricanes. These were carefully concealed within camouflaged 'hides', with a decoy airstrip and fake planes close by. Semi-circular anti-aircraft positions comprising four large dug-outs were easily identifiable from the air, and although at the time they represented Tobruk's only real air defence system, it was deemed impractical to camouflage them.

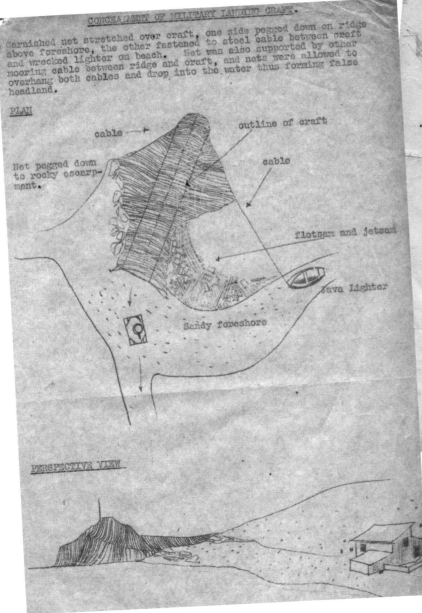

CONCEALMENT OF MILITARY LANDING CRAFT.

Garnished net stretched over craft, one side pegged down on ridge above foreshore, the other fastened to steel cable between craft and wrecked lighter on beach. Net was also supported by other mooring cable between ridge and craft, and nets were allowed to overhang both cables and drop into the water thus forming false headland.

PLAN

cable →

outline of craft

Net pegged down to rocky escarpment.

cable

flotsam and jetsam

Java Lighter

Sandy foreshore

PERSPECTIVE VIEW

P.S. Salvage actively with U/S MT.

To C.R.E,
TOBRUK Div.

The above sketch shows the Camouflage treatment for the Distillery, which I hope has not been hit. After the attack this evening it seems that now is the time to take action on this scheme. The German recce planes will probably be over tomorrow morning so it is suggested that at least the hole be simulated at once, using oil or tar. Other details as follows:-

A. Dummy crater to be dug 18" in depth, further depth to be suggested by darkening hole on SOUTH side with oil or tar

B. Grey colouration caused by bomb in vicinity of building by use of cement fondu or sprinkling of oil through perforated can.

C. Debris to be scattered about.

D. Oily rags to be burnt in any adjacent u/s building.

E. From a vertical view it is difficult to see the small chimneys so it might be a good idea to lay 8 oil drums end to end across roof.

Above: William Murray Dixon's plans for the bogus devastation of Tobruk's water distillery.
Property of W. M. Dixon, ARCA

Left: Instructions for concealing military landing craft in Tobruk harbour.
Property of W. M. Dixon, ARCA

Further deceptive measures were therefore brought into play. Nearby sites were converted into two or three dummy anti-aircraft units, complete with dummy troops and guns created from scrap material. These decoys were brought to life during air-raids by exploding Italian grenades. As the siege developed, additional confusion was heightened by the replacement of these dummy sites with live guns. The whole operation was declared to have been very successful, and not one gun was lost in action.

During preparations for Operation Crusader, the offensive planned by General Auchinleck to overthrow Rommel's troops, it became clear that the important almost-completed railhead at Misheifa was the prime candidate for attack and therefore camouflage. The job of arranging this enormous and seemingly unfeasible task fell to Stephen Sykes, who decided that as it was impossible to hide the railhead, only a decoy to draw attention away from it would work. The railhead was being built by the New Zealand Railway Construction Company, which was ready to join in with the deception plan.[13] The idea was to construct a dummy railhead at the end of a six-mile 'extension' of track, ostensibly for the delivery of tanks. This would be built up to resemble a depot with sidings and ramps, complete with fake engines and trucks which could be moved around at night. The construction company's involvement with the project was hindered by its work on the actual railhead. Track ran out, and dummy rails had to be improvised with beaten tin cans. Rolling stock was created against all odds in the midst of desert

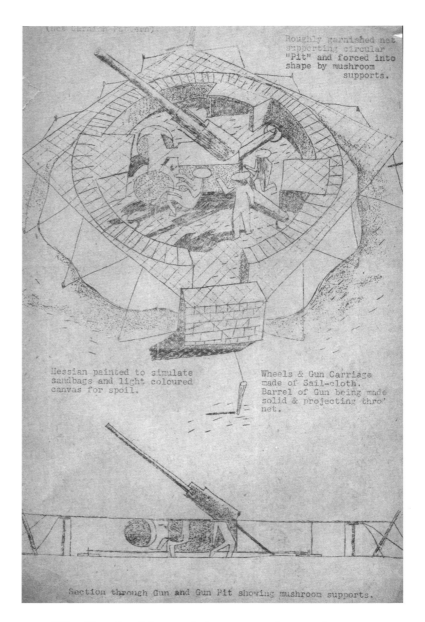

William Murray Dixon's depiction of a dummy anti-aircraft gun pit, with dummy operators.
Property of W. M. Dixon, ARCA

storms from sticks and canvas, timber when it was available, and palm fencing along with any scrap material that could be mustered. The scale of the vehicles had to be reduced to two-thirds of their original size, owing to the shortage of materials. The whole job seemed beset by obstacles, but eventually Sykes and his team created around 50 dummy wagons and trucks, one locomotive complete with smoke (courtesy of an old stove), and several tanks: 'dummy men are grubbing in dummy swill-tanks, and dummy lorries are unloading dummy tanks, while a dummy engine

116

Brian Robb, Laying Track for the Dummy Railway.
A reduced-scale train can be seen in the background.
Geoffrey Barkas, The Camouflage Story, *Cassell plc, a division of the Orion Publishing Group (London), 1952*

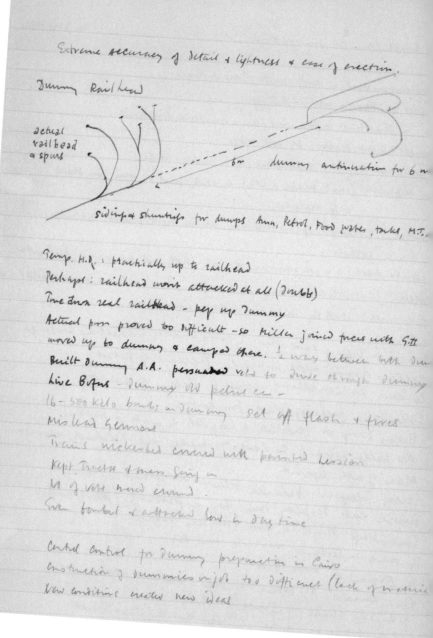

Ashley Havinden's notes from Jack Miller's lecture on desert camouflage.
Imperial War Museum

puffs dummy smoke in the eyes of a possible enemy'.[14] A sandstorm on the final night of construction nearly destroyed the most crucial areas, but the depot was finally ready for attack, and bombed it was – extensively, a number of times, therefore protecting Misheifa from major damage.[15] As concluded in Ashley Havinden's notes: 'Railway: with dummy railhead – v. successful. They bombed it.'

Despite the camouflage team's remarkable efforts, and the tireless support of the British Forces and their allies, Tobruk finally fell in June 1942. The British Army, together with the New Zealanders' XIII Corps, retreated eastwards towards Egypt in the belief that this was the last remaining place where a stronghold could be mounted west of the Nile Delta, the Egyptian border and Suez. Should the Germans conquer El Alamein, it would be easy for their army to then take possession of Alexandria, the Nile Delta, Cairo, Port Said and the Suez Canal (which had seemingly been protected throughout this time by a deceptive lighting system devised by Jasper Maskelyne). On 8 July Rommel's army made an abortive attempt to break through the defences, and then things went ominously quiet. Both sides began an intense period of preparation. Under the direction of General Auchinleck the British troops began to develop a plan to trick the enemy into thinking that a sizeable army was amassing behind the battle lines. While genuine supplies were brought in to the northern line of attack to El Alamein, dummy encampments were set up as decoys to the east, complete with track networks, food and ammunition dumps, vehicles and personnel.

These operations were designed to pave the way for the long-awaited *coup*. This was the brainchild of Colonel Barkas, who was supported by the painter Tony Ayrton, drafted from military camouflage duties in Abyssinia, and Lieutenant Brian Robb, later to become head of illustration at the RCA. Meanwhile, Rommel was observing the 'build-up' of British power and slowly amassing more troops to his beleaguered army, and the longer this took, the greater the advantage to the British camouflage team in the construction of a complete dummy offensive.

Brian Robb, Little-Known Units of the Western Desert – Rapid Disorganisation Group. *Brian Robb, from* My Middle East Campaigns, *Collins, 1944, reproduced with permission of Quentin Blake*

117

*William Murray Dixon,
Coming Home on Ship,
c.1943. The inexhaustible
Dixon continued to record
the events of the Second
World War.*
Property of W. M. Dixon, ARCA

Dummy soldiers in the desert.

dummies. Live guns would then be substituted for them, and undisguised tanks which had deceptively been set amongst the dummies would also go into action. In the north, a large concentration of lorries were replaced the night before battle by tanks disguised as the same lorries. Where the tanks had been resting, a battalion of dummy tanks would replace them, looking ready to move south. Peter Proud organised the digging of a dummy pipeline (trenches filled with a line of empty oil-drums, which were repeatedly used as the trench was extended and filled in every day), eventually giving up the arduous task of digging to replace it with a skilful yet realistic image spray-painted on to the sand in used sump-oil. Dummy food and ammunition dumps were also built up slowly to suggest a delayed date for assault.

119

William Murray Dixon's 'small foolscap drawings' which were sent from Tobruk to Cairo encapsulated the programme for the delivery of El Alamein. He was then drafted to Baghdad to begin work on the plans. An enormous dummy army, with 'vehicles and dummy soldiers – overalls supported by a post, with a tin hat' would replace the actual army. In the southern area, dummy artillery and tanks were installed as a false threat, and the enemy would be allowed to identify them as

This complicated combination of concealment and display was so carefully planned and so finely tuned down to the last detail that it worked. Montgomery's 8th Army moved into attack position on 23 October 1942. Vicious and prolonged fighting ensued, and it was not until 4 November that the British finally emerged, exhausted but victorious from the Battle of El Alamein. A remarkable plan of deception and counter-deception – 'sinister games of bluff' – had led to triumph in this stage of the war.[16]

CHAPTER EIGHT

Operation Overlord

As the war continued, it became increasingly obvious that the only real way forward for the creative camouflage team was to learn from the desert experience, and explore the arts of illusion and trickery more widely. The outwitting of the enemy and a successful invasion of occupied Europe would only be possible by means of a meticulously-planned operation based on pretence and misrepresentation.

This was where the imagination and expertise of the development team at Farnham really came into its own. In addition to being the major camouflage training centre, Farnham was to come up with new inventions and developments to be used for every imaginable camouflage purpose. According to David Medd, who eventually became head of the workshop, the CD&TC's direct representation in the War Office enabled it to be aware of army movements for the year ahead. This was essential in the preparation of appropriate concealment schemes. Medd specialised in acoustic and electrical work (for example, lights and loudspeakers), but his output included some seminal devices such as the dummy landing craft designed by Basil Spence and

Opposite: Demonstration of Farnham's featherweight 'Device 91'.
Property of David Medd

Below: A rubber tank at Dunlop's headquarters, and the same tank stowed in a bag.
Property of David Medd

122

FOREWORD

This handbook presents, in skeleton form only, the basis on which Camouflage displays should be built. As with a diagrammatic drawing in an anatomy book, this booklet also shows the figure in one position only and is a collection of the bones and muscles of visual deception without the flesh ; this must be added in every case. Further, the figure is presented in one aspect alone without any changes in position of arms, legs and head, without any bend in the body, or twist to the torso. Just as every human or animal body has a thousand different postures, from variations of individual fingers or toes to alterations to the whole structure, so too on this skeleton of equipment and associated features, each and every display must vary and individuality and liveliness be obtained.

Just as in the clothed human body, or in a feathered bird, much may be hidden, so too in display, on appropriate backgrounds, much may be assumed as hidden, and only a part shown. Yet that part must suggest the rest to which it belongs and must be co-ordinated to the whole, at least in the mind of those staging the display.

Just as a painter, using his knowledge of anatomy, can tell a story by his painting so too must a story of battle be told in camouflage display.

The anatomy book and this booklet shows in detail a single figure only. The artist frequently uses many figures in a single picture, and in Camouflage the display scene must be built up with many individual signatures.

As in painting it is the main lines of the composition as a whole that makes the work great or little, so in display it is the composition of the associated features of track and spoil, of smoke and bivouac, over a considerable area, that makes it significant or negligible. The individual dummy devices are no more important than the details of costume or accoutrement.

The flesh, that is associated features, must always be presented. The clothes, that are the devices, may at times be scanty or even non-existent.

C.D. & T.C., RE., FARNHAM

The elegantly-phrased foreword to the Technical Handbook on Camouflage Display, *probably written by James Gardner.*
Property of David Medd

used to such splendid effect during D-Day operations.

The Farnham team continued to include many artists and designers who were able to bring their pre-war expertise into play. One of the most important resources was the designers' knowledge of industrial contractors who would be able to develop experimental techniques for wartime use. David Medd recalls three major factories who turned their production around for camouflage purposes. The Bradford weavers Listers Mills used their velvet looms to create a two-inch pile fabric: the two layers, which would normally be cut apart, were left uncut and coated with rubber. The resulting fabric became known as 'slab' and was used in camouflage schemes where large flexible flat surfaces were required. Dunlop's giant tyre factory in Birmingham provided the rubber coating, and also produced scores of inflatable rubber dummy tanks, lorries and later, landing craft. If these inflatables needed a supporting structure, Cox & Company of Watford, metal fabricators, could manufacture aluminium frames.

An excerpt from the 21 Army Group's War Diaries, written on New Year's Day 1944, defines camouflage as 'the art of concealing from the enemy the disposition of men and materials, and also of indicating false positions by means such as the use of dummy equipment'.[1] The diaries go on to explain that while until recently the term 'camouflage' had been taken to mean both concealment and display, the use of display (rechristened 'indication') now carried optimum importance and would rely on special intelligence outside the

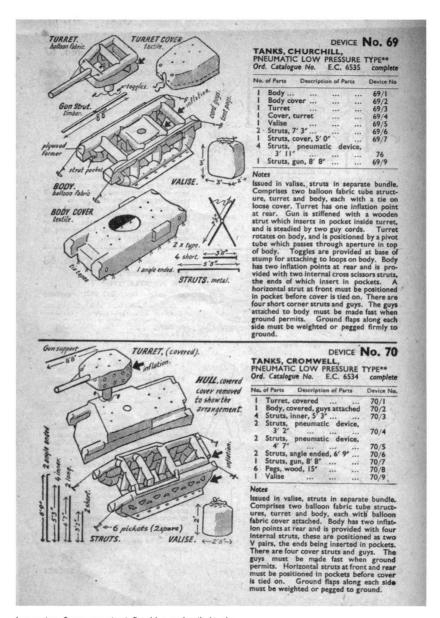

Instructions for constructing inflatables, as detailed in the Technical Handbook on Camouflage Display.
Property of David Medd

123

scope of the normal camouflage service.[2] This was not to say that former camouflage methods in the form of concealment, disguise and decoy had become obsolete; these techniques would have to be combined with new ways of tricking the enemy into making crucial mistaken assumptions.

The newly created 21 Army Group Camouflage Service modelled itself on the now disbanded camouflage department at GHQ Middle East (Cairo). The Group's army list for May 1944 names Major G. B. Money-Coutts, Captain W. V. Cole and Lieutenant V. A. Stanley-Adams as its officers. In command of the Camouflage Pool was Major John Hutton. This small group of advisers at HQ Southwick Park was part of the structure of a special unit, G(R), in which all deceptive operations were put under the control of the rather impulsive Lieutenant-Colonel D. I. Strangeways, who had been General Alexander's deception officer in North Africa. Camouflage troops would be 'under the command of a small mobile HQ' and would be responsible for 'all forms of tactical deception'.[3]

From December 1943 onwards staff began to leave England to report for camouflage duties on the Normandy beaches in preparation for the Allied invasion. One staff captain (camouflage) was put in charge of each 'Sub-Area (Beach)', and each Beach Group was staffed by a captain and sergeant (camouflage) and a 'Driver (Transport, Jeep 1, as soon as possible)'.[4] There was often an intelligence officer attached to the group. Beach Groups were put in place at an early stage so that all personnel (officers and troops from the Pool)

could become used to the new terrain and treat it in suitable ways. The first operation would commence with the arrival and concealment of stores and equipment, obviously followed by vehicles, artillery and personnel. In addition to John Hutton, a familiar name from North Africa was that of Captain Stephen Sykes who was posted to 21 Army Group Camouflage Pool No. 5 Beach Group with effect from March 1944.

The complex and daring sequence of strategies developed to convince Hitler that the British would invade France at its most accessible northern point was code-named Operation Overlord. A gigantic plan of deception, Bodyguard, was set up by the London Controlling Section (Britain's secret war intelligence agency) to make the enemy believe that it had obtained accurate inside information that the Pas de Calais would be the point of invasion. There were three main threads to this plot. Firstly, it was necessary to give the impression that the attack could take place in any one of the occupied countries in Northern Europe, so that German troops were drafted away from the real action. Secondly, once it was clear to enemy intelligence that a cross-Channel invasion would take place, it would be deceived into thinking this would happen north of Normandy at the narrowest point of the Channel; and thirdly, it was important to convince the enemy that the timing of the attack would be later than actually scheduled.

Operation Fortitude North was launched to lay the trail of deception. Over the coming months, the enemy was able to intercept cryptic radio information giving

124

*Aerial photograph of
dummy tank landing craft,
and a record of one under
construction.*
Property of David Medd

125

clues about a Northern invasion plan. A totally fictitious
Allied '4th Army' was seen to be amassing along the
east coast of Scotland, evidently preparing for battle.
Dummy landing craft, weapons, tanks and troops were
assembling, backed up by bogus radio information and
a whole network of deceptive data aimed at leading the
enemy to believe that Norway was about to be invaded,
so rendering Germany's northern defences highly
vulnerable. By early 1944 18 German divisions had
been drafted to Norway and Denmark in readiness for

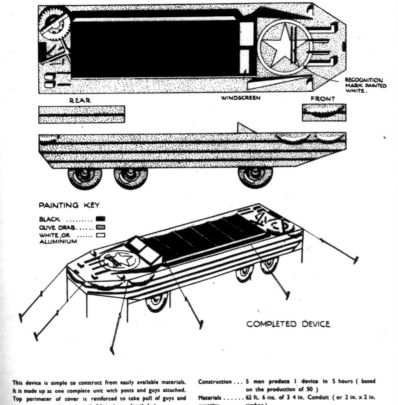

REAR   WINDSCREEN   FRONT

RECOGNITION
MARK PAINTED
WHITE.

PAINTING KEY

BLACK ....
OLIVE DRAB......
WHITE, OR ....
ALUMINIUM

COMPLETED DEVICE

This device is simple to construct from easily available materials. It is made up as one complete unit with posts and guys attached.
Top perimeter of cover is reinforced to take pull of guys and prevent posts pushing through fabric (see detail A)
Guys and pickets at each end require to be strong and long, as these take main strain and keep craft in shape.
Well of craft is a swag of black material, lightweight cords connect two sides of gap to prevent distortion.
Windscreen is improved by application of aluminium paint or light reflective material. Painting should be kept simple and bold.
If a number are to be displayed, well of craft should be varied by patches or paint representing load. Some may with advantage be hooded, for dimensions of hood see sheet ?

Construction ... 5 men produce 1 device in 5 hours (based on the production of 50)
Materials ...... 62 ft. 6 ins. of 3 4 in. Conduit (or 2 in. x 2 in.
quantity   timber)
and type   3 ft. 8 ins. of 5 16 in. M. S. R.
This list   108 ft. of 1 in. & 16 ft. of 1 2 in. Cordage.
includes   50 yds. (approx.) of Fabric (8 oz. Hessian, or
material   similar material)
for making   2 1 2 gals. (approx.) of Paint (Olive Drab
pickets   White, Black)
and cleats   5 ft. of 1 2 in. x 1 2 in. Timber.
Loading .... 75 per 3 ton G. S. Lorry Weight .. 80 - 85 lbs.
Erection Time ...4 men - 2 minutes.

Diagram of a dummy amphibious truck, or 'duck', reproduced from the Technical Handbook on Camouflage Display.
Property of David Medd

an invasion by the giant '4th Army', thereby seriously depleting Hitler's army further south.[5] A 'top secret' document from the Air Ministry records that 'the cover plan in Scotland seems to have been most successful'.[6]

To balance this illusion, another diversion in the shape of Fortitude South rolled into action. In late summer 1943 a 'First US Army Group' (FUSAG) was understood to be mobilising in southern England, ready for battle in the Pas de Calais, and again supported by leaked information and fake radio signals. This operation was mounted at Southern Command where, under the direction of Major Edward Seago, tented and hutted camps with all necessary equipment were assembled to accommodate very visible American troops. Backing up the US Army hoax was a programme of deception situated around the rivers and channels of the south-eastern coast of England, designed to protect the whole of the south coast and deceive the enemy into believing that craft were amassing in the waterways of Essex and Suffolk. Naval staff were responsible for manning the protective decoys in the Solent and around Southampton. Despite setbacks such as the need to skirt mined areas, and unladylike pedigree cows which trampled on vital equipment, dummy tank-landing craft (as used in North Africa) were surreptitiously moored in the discreet creeks of East Anglia 'ready for attack', with the illusion of pre-battle activity to back up their existence.[7]

The design for these craft is attributed to Captain Basil Spence, who was attached to Eastern Command HQ. Working to sketches and blueprints provided by

126

Left to right: Ashley Havinden, Eugene Mollo, Ella Mollo and Margaret Havinden.

Property of Michael Havinden

at night.'[10] Strangeways seems to have frequently upset official camouflage arrangements by attempting to overrule and to exert control. However, he and his team successfully joined forces with Colonel Turner's trained troops to provide vital lighting protection on the beaches.

A supply of fuel would be necessary to support the planned invasion once troops had reached Normandy. The engineer working on the design and construction of this astonishing pipeline under the ocean (its acronym wittily providing the operation's codename Pluto) was Russian-born Eugene Mollo, a Royal Engineer trained at Farnham. The Disney connection inspired codenames for the various components of this system. There were two planned petrol terminals: the first at Dungeness, Kent, was christened Dumbo, and the second at Sandown Bay, Isle of Wight, Bambi. An eight-inch pipe between the two terminals was given the name Solo, and the reserve storage tank was called Toto. Having been a staff lieutenant at HQ Hants and Dorset District (and previously at Darlington) Captain Ashley Havinden was called up in August 1943 for camouflage duty at Ryde, Isle of Wight. On the strength of his achievements in the planning of camouflage for Bambi, he was subsequently made chief camouflage officer for Bambi, Dumbo and Solo, 'attached' and based at the War Office in Whitehall. To give an idea of the vast scale of Operation Pluto, Dumbo alone consisted of 30 thirty-foot-long storage tanks served by 30 diesel pumps, and continuing as 12 submarine cables. Bambi was a slightly reduced version of this.

127

Spence, the Shepperton Studios team built 225 fake 'Bigbobs' (LCTs or landing craft for tanks) and 138 smaller 'Wetbobs' (LCAs, otherwise known as troop landing craft for assault).[8] According to James Gardner, the structures consisted of an aluminium canvas-covered frame floated on oil-drums, with launching-wheels sequestered from shot-down German aircraft.[9] Colonel Turner points out that Lieutenant-Colonel Strangeways rather foolishly had these craft constructed in full view of the town of Yarmouth, therefore jeopardising decoy activities in general: 'I do not feel that the display in East Anglia will have any appreciable success … It is a pity, because we took a great deal of trouble over our decoys, and went so far as to put lights on some of the Bigbobs so as to marry them up with the shore lighting

128

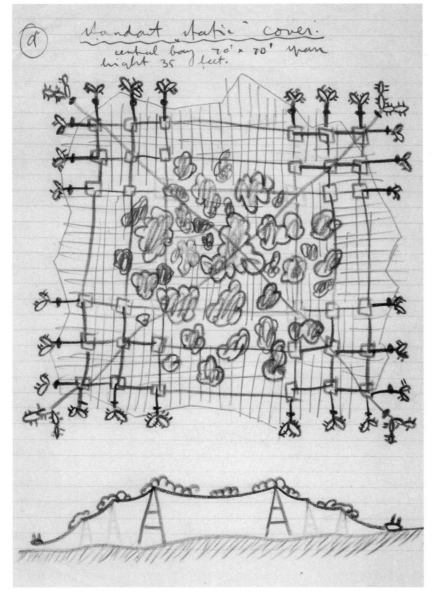

Coloured sketch, probably by Eugene Mollo, Camouflage Cover
for Fuel Tank, *1944.*
Imperial War Museum

Together Havinden and Mollo conceived an audacious design for the camouflage of a supply facility which would consist of the carefully hidden tanks, disguised by Mollo's construction of metal tripods and steel-wire ropes holding up a cover 'dressed with 3-dimensional features representing scrub, bushes and treetops etc.'.[11] Troops were trained in daytime to erect the framework under cover of darkness. From these terminals, petrol pumps (some installed in holiday cottages in Dungeness) would relay the fuel via shoreline reservoirs to the supply pipes running underwater to France.

In order to divert enemy intelligence away from this lifeline, at the request of the London Controlling Section, Basil Spence designed a giant decoy oil storage facility with the assistance of the Shepperton technicians. The installation took up three square miles of the Dover coastline, and consisted of everything a genuine oil terminal would have – storage tanks, vehicle bays, pumps, pipelines, troops and anti-aircraft artillery. To create a convincing illusion, it was officially 'opened' in the presence of King George VI and General Sir Bernard Law Montgomery.[12] The whole bogus operation was remarkably successful, with British fighter planes keeping enemy reconnaissance at bay throughout construction and afterwards.

It was perceived that the capture of an appropriate French port from which to stage the Normandy landings would be extremely difficult, and therefore the decision was taken to construct giant floating harbours, complete with quays and depots, which

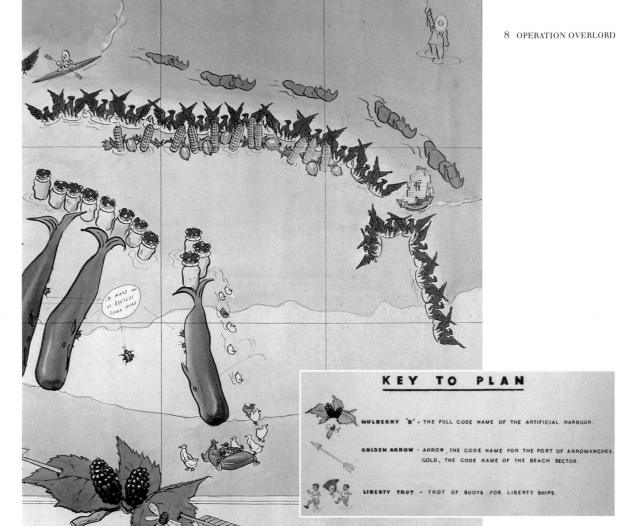

KEY TO PLAN

MULBERRY 'B' · THE FULL CODE NAME OF THE ARTIFICIAL HARBOUR.

GOLDEN ARROW · ARROW, THE CODE NAME FOR THE PORT OF ARROMANCHES.
GOLD, THE CODE NAME OF THE BEACH SECTOR.

LIBERTY TROT · TROT OF BUOYS FOR LIBERTY SHIPS.

LIBERTY SHIPS ANCHORAGE ·

RHINOS · THE POWERDRIVEN PONTOONS ON WHICH CARGO WAS BROUGHT ASHORE.

Detail of the Mulberry harbour cartoon drawn by Francis Marshall at the Admiralty camouflage headquarters in Bath.
Property of Julia Atkinson

could be used to supply the attacking Allied troops. Designed in Bath at the Admiralty's temporary head-quarters, two huge 'Mulberries' were built from approximately 2 million tonnes of concrete and steel, surrounded by floating roadways supported by pontoons. The superstructure was protected by a barrier of specially-made breakwaters, with the addition of the hulks of 70 scuttled ships (called 'Gooseberries'). The

different components of these man-made harbours were built in various parts of Britain, and secretly submerged off Selsey and Dungeness, on the south coast. Following D-Day on 6 June 1944, the Mulberries were assembled and towed to their destinations at St Laurent and Arromanches in Normandy. The harbour at St Laurent was subsequently damaged during a storm, leaving the Arromanches-based Mulberry to take over all delivery and storage duties. According to US Captain Manley Power, it 'worked like a dream and made a really remarkable contribution to efficiency'.[15]

Operation Neptune was launched on the night of 5 June 1944, helped by a short let-up in perilously stormy weather. Thousands of Allied troops, their weapons, vehicles and equipment, sailed for Normandy supported by an enormous squadron of fighter planes and destroyers. Lancaster bombers flew in a ship's convoy formation north of the armada to drop 'chaff' (this consisted of strips of aluminium foil in various sizes, and was given the codename Window) which would create an echo in enemy receivers and imply that there was a large flotilla heading for the Pas de Calais coast. Balloons and radar reflectors operated by the Navy also helped the illusion to succeed; dummy parachutists were dropped by the RAF over the north coast of France. With enemy forces sent by sea and air to interrupt the imaginary fleet, the real task force was left free from attack, and thus began the fabled and almost unbelievable Normandy landings.[14] The German 7th Army were caught off-guard, and despite strong resistance, the Allies managed to fight their way ashore and establish supremacy along the northern French coastline.

The beach areas where the invasion started were named Utah, Omaha, Juno, Gold, and Sword. In the continued advance into France subsequent to Operation Neptune, when much of the German Army was still awaiting the Pas de Calais invasion, camouflage measures and training continued. Instructions from the War Office advised that each Beach Group was to ensure that camouflage arrangements would continue to be understood and implemented. Camouflage officers had to check that troops appreciated the measures taken for the protection of stores, equipment and themselves. Basil Spence found himself responsible for Sword beach and any camouflage necessary for it: 'It was decided not to continue scheme except for Smoke Screen in meantime. Air photographs of Beach area were given to Capt. Spence, who gave talks to units … illustrated by air photos'. Stephen Sykes, meanwhile, attached to 'No. 5 Beach Group' was working on 'close quarters camouflage' analysing the use of improvised sniper suits and dummy figures.[15] Major John Hutton's job was to enable continual liaison between Pool officers, and to oversee the concealment of certain armoured divisions prior to attack. Nobody was under any illusion that the Normandy invasion meant a speedy end to the war – War Office diaries caution that 'monthly reports will be submitted on the progress of Camouflage training', suggesting a resignation to the elapse of a considerable length of time before the cessation of hostilities.

Sergeant Edwin La Dell, subsequently head of the department of printmaking at the RCA, is also mentioned for post D-Day camouflage duties at HQ Southwick Park. The War Diaries of 1944 record official activities down to the smallest movements of a unit involved in the details of war:

28 June 11.00   Sgt. La Dell to Portsmouth to obtain some coloured inks in order to colour French maps.

Robin Darwin: T. E. La Dell, 1957. Oil on canvas.
Royal College of Art Collection

28 June 14.45   Sgt. La Dell returned from Portsmouth.
3 July   Sgt. La Dell to Didcot to collect photo-
                graphic stores.
19 July   Sgt. La Dell left for overseas.[16]

It is evident from this excerpt that preparatory work for the liberation of France began when the 21 Army Group Camouflage Service was based at its UK head-quarters. In mid-July (well after the Normandy landings) the unit moved to Le Tronquay, inland from the French coastline, where camouflage ideas contin-ued to be developed in order to aid the continued assault on enemy troops. On 5 August 1944, for exam-ple, the War Diaries record that Major Money-Coutts and Major Hutton visited the 21 Transport Column at Luc-sur-Mer, where they were asked if Captain Spence could be attached to this unit to train new drivers. This was agreed. In September, Hutton visited Spence at this division, where his work was coming to an end and arranged his transfer to 'undertake investigation of the Boulogne area'.[17] Spence became Hutton's deputy at Farnham on their return to Britain and the winding-down of the CD&TC, where amongst other techniques he gave the Army instruction in snow camouflage (which he learned early in the war with the Commandos in his native Scotland). The combination of Spence's ground-breaking architecture and Hutton's emotional engraved glass in their collaboration on the post-war rebuilding of Coventry Cathedral serve well to illustrate the lasting alliances forged in the service of camouflage.

131

CHAPTER NINE

Admiralty Camouflage

132

AT THE START OF THE WAR it appears that there was some confusion over the subject of camouflage for Royal Navy ships. The Admiralty had concluded in 1937 that the dazzle-painted ships of 1914–18 had been spectacular but not successful, except for having a positive effect on the morale of their crews. The majority of HM ships between the wars were painted in a rather more sober livery: plain dark grey for the Home and Reserve fleets, and lighter grey for large vessels in the Mediterranean and further afield. Craft stationed in the tropics were white or two-tone with light grey 'upper works'.[1] Merchant ship crews were allocated no particular instructions other than that their vessels should not be painted grey, for fear of being confused with warships, and attacked.

In 1939 the Admiralty set up a continuous supply convoy to and from Africa. The Atlantic Ocean became the graveyard for an unbelievable number of British craft, destroyed by enemy raiders and U-boats. The superiority of the British Navy, supported by its escort aircraft carriers (converted cargo ships) caused the enemy to rethink tactics and to resort to attacking British merchant ships delivering food and essential supplies. This submarine devastation continued throughout 1940 until the breaking of the Enigma codes in early 1941. At this point the Home Office issued a command that 'from now on all HM ships should be camouflaged and … the subject kept under review to produce the best designs'.

Although dazzle painting was not recommended, an early instruction to treat ships with 'disruptive' patterns had been issued in late 1939 by Admiral Dunbar-Nasmyth. He ordered that the destroyers *Grenville* and *Grenade* should be painted with experimental patterns to break up their outlines: the former in a two-tone grey geometric pattern, the latter with softer patches in two greys and a neutral buff colour.[2] Many disruptive designs and colour schemes were tested over the following two years, resulting in a report from the Admiralty stating 'evidence that … light or dark disruptive schemes reduced sighting range was inconclusive … The secondary effects – confusion of inclination and disguise of type and class of ship – were not marked'.[3] However, the proposals researched and designed by Lieutenant Peter Scott of the Royal Naval Volunteer Reserve, which were based on his

Opposite: Claude Muncaster's painting of a ship in dazzle camouflage, 1942. Watercolour on paper. RAF Museum

knowledge of camouflage in nature, were later officially approved and tested on his ship HMS *Broke*. In late
1941 Scott developed his 'Western Approaches' scheme,
named after the rectangular area of the Atlantic, lying
west of the British Isles and extending north to Iceland,
where the Battle of the Atlantic was so viciously fought
throughout the war. This was based on white with
patterns of pale grey and blue, and became the standard
colour scheme for coastal forces and smaller Royal
Navy ships in the North Atlantic.[4]

*Opposite: Eric Ravilious,
Leaving Scapa Flow,
1940, shows the use of
strong disruptive
geometric patterning on
the ship's funnels. The
vessel is probably HMS
Highlander. Watercolour
on paper.
Cartwright Hall Art Gallery,
Bradford*

In October 1941 the Admiralty stated that the
primary aim of camouflage was 'to reduce sighting
range to the minimum'.[5] When devising wartime camouflage, ships must be the most complex and difficult
subjects for treatment; there are so many influences on
their perception, such as inclination and direction, and
the ever-changing conditions of atmosphere and light at
sea. An ongoing debate revolved around whether pattern
was effective in reducing range of visibility, or if the
reverse was true. Pattern worked within a certain viewing range, when the perception of outline became
distorted and confused. Beyond this range, a ship 'will
always appear as a uniform tone whatever the lighting
conditions … the nearer this tone matches the tone of
the sea background the less visible the ship will be'.[6]

The problems to be solved by the naval camoufleur
were comparable to, yet vastly different from, those
which beset the designer of static camouflage: 'it is by
no means so easy at sea. The background … is never of
a constant tone.'[7] It became evident that a research and
development department would be necessary in order
to investigate the endless factors affecting the disguise
and concealment of the Fleet, and for this reason the
Admiralty decided to create a naval camouflage unit
alongside the Civil Defence Camouflage Establishment
at Leamington Spa.

For the duration of the war, the Admiralty, while
maintaining its Whitehall headquarters, evacuated
around 4,000 staff to the city of Bath. It was considered
essential that Admiralty camouflage officers should
have a 'knowledge of seafaring methods and practices';
consequently, naval officers who had previously
received a training in art or design like those in civil
camouflage were recruited.[8] Ex-RCA student Leslie
Atkinson and the fashion illustrator Francis Marshall
were members of the Bath organisation. They had
joined the Navy early in the war and were subsequently selected to leave their training ship HMS
Ganges to report for camouflage duties. The marine
painter Oliver Grahame Hall (known professionally
as Claude Muncaster), who was attached to the
Training and Staff Duties Division of the Admiralty's
Directorate of Scientific Research, was asked to oversee a team of design staff mainly seconded from the
Royal Naval Volunteer Reserve (otherwise known as
the Wavy Navy on account of the undulating gold
stripes on the sleeves of its officers). He was assisted
by Lieutenant David Pye and Admiralty Liaison
Officers Robert Goodden and R. D. (Dick) Russell
(both architects in peacetime), who had completed a
naval training course at the Royal Naval College,
Greenwich. Stephen Bone, who was also recruited as

135

Telephone : Leamington Spa 2190

Telegrams : Camest, Leamington Spa.

Correspondence on the subject of this letter should be addressed to the Director, and the following number quoted :

MINISTRY OF HOME SECURITY,

DIRECTORATE OF CAMOUFLAGE

REGENT HOTEL, LEAMINGTON SPA

David Pye's detailed notes and sketches of sea conditions.
The National Archives

Nos. 25 and 26 (Diagram and Sketch)

Light M.S.S. matching sky in S.B.R. 12+

Time 101700

Position 58° 30' N, 27° 00' W.

Height of Eye 43 feet

S.B.R. 12 +

Wind (N. 4)

Sea and Swell (22)

Visibility (Very good)

Cloud Perfectly clear sky up sun down sun pole pink-grey cumulus cloud background to the Light M.S.S. ship. Background to the corvette as sketch

Brightness Light M.S.S. ship "a perfect match to sky". Brightness of corvette remained constant but brightness of background changed.

Remarks The corvette was hull down and "changed while still sunlit from too light, to a perfect match (for the sky) as she passed from a dark part of the background squall (D of sketch) to a light part (L of sketch).

Inference The type of cloud background largely determines the effect of camouflage in high S.B.Rs. when viewed from the surface.

137

a camouflage officer, was later elected to the status of official war artist to the Navy.

The Admiralty camouflage headquarters at Leamington were installed in the Municipal Art Gallery, and Hall remembers meeting Lieutenant-Commander Yunge-Bateman, Director of Naval Camouflage, who was responsible for a staff comprising at that time three camouflage officers, five senior and three junior technical assistants, and a modelmaker.[9] Their brief was 'to state the requirements for the painting of every seagoing ship and vessel of the Fleet' and to constantly analyse all factors which made easy targets of ships.

There were two major steps in the preparation of a naval camouflage design. The first one was to find out as much as possible about the relevant craft's posting and potential navigational conditions as seen from the air. Frequent reports would be made by camouflage officers on flights around the coast of England and Scotland. A series of David Pye's reports from around 120 observational flights during May and June 1943, accompanied by drawings and photos, shows the minute detail recorded. Another set logged by Major Alan Durst, Royal Marines (elected Admiralty Representative on the Camouflage Committee), reveals the artist's lyrical sensibility to atmospheric conditions: 'Colour of sky. Blue in clear. Warm almost pink pearly clouds. Colour of sea. Bluer and a shade darker than sky at horizon. Grey with pink tinge closer.'

According to Robert Goodden, the officers 'didn't always come up with the right solution' for a camouflage scheme. Even with results, for example, from 'seven observations made by each of four observers on each of five ships', no definitive conclusions were drawn.[10] It must have been a particularly difficult task to

select a suitable average from the conditions recorded.

The watercolourist Lieutenant Donald Currie, an ex-Royal Navy officer from the Scientific Research department attached to the Admiralty, undertook aerial observations using Peter Scott's ship HMS *Broke* as a base, while on a reconnaissance voyage in January 1942 from Western Approaches HQ in Liverpool to Iceland and back. His recordings show that disruptive camouflage had little effect on average visibility in the north Atlantic (range 12–15 miles), and as long as ships were grey or another pale colour they were not obvious in reasonable light.[11] Richard Guyatt, based in Leamington with the civil camouflage team, remembers being asked to accompany Currie on a series of Sunderland bomber flights over the Atlantic. Their aim, in Guyatt's words, was to 'discover the colour of the sea' with a view to finding the perfect tone for ships' camouflage. This was carried out by making 'half-hourly daytime pastel sketches', the average colour resulting as a grey called 'S100', which was eventually used throughout the merchant navy. Remarkably, when analysed, it turned out to be only 0.01 per cent different from the colour the Americans had used for their fleet in the First World War.

The second stage in creating designs for naval camouflage was to construct an accurate model of the vessel in question and to put it through a series of simulated tests. For this purpose a large tank was constructed at Leamington. It was designed by the marine artist Frank Mason, and based on a similar one he had made for testing dazzle effects during the prev-

ious war.[12] An Admiralty report of 1942 mentions a tank, probably replacing Mason's, which had been developed by 'an officer who had worked as a film director' – possibly the designer Wilfred Shingleton. This tank was 'about the size of a room, with model ships 12–24 feet long'.[13]

Felicity Fisher, one of the junior technical assistants (JTAs) in Yunge-Bateman's team, remembers that her main task was to enlarge, with extreme accuracy, plans and elevations of battleships illustrated in *Jane's Fighting Ships*. Scale drawings were made with a steel pen producing a regular heavy line on treated fine cotton, and the image was then reproduced by a photographic process on to special paper. (Fisher recalls that pieces of fabric discarded because of errors could be 'washed and transformed into useful garments'!) Next, the modelmaker (a professional cabinetmaker by trade) translated the scale drawings into wooden models and delivered them to a camouflage officer who would develop a design ('elaborate patterns of problems' according to Robert Goodden).

Fisher and Victorine Foot, another young technical assistant, recall that the camouflage officers, glamorous visitors in their naval uniforms, then viewed the model ships on the tank's surface through 'back-to-front binoculars from a box darkened by blackout material'. The tank's simulated lighting and physical conditions were operated from a switchboard with labels such as 'grey day', 'bright sun', 'moonlight', 'moonless' and so on. A light-hearted description of the viewing room appeared in *The Fortnightly DO* in spring

Opposite: The naval viewing tank at Leamington. A backdrop is being installed in order to test model ships in simulated conditions. Reproduced with permission of Getty Images

140

Frank Mason, Modelmaker's Shop, Directorate of Camouflage, (Naval Section), *Leamington Spa,*
purchased in 1943. A detailed and atmospheric image of a skilled cabinetmaker's wartime work.
Imperial War Museum

'The Arctic comes to Lilliput'. Lt-Commander Yunge-Bateman tests a model battleship in the viewing tank.

1942: 'a large shallow tank is arranged with fans to ripple the water, all most realistically, and a large twenty-foot mirror that reflects the real clouds at the back. Civil servants are hired to blow smoke through their noses and, seen … through a viewer one might just fancy oneself just passing off the North Foreland. Another gigantic peepshow produces a tropical storm by pressing a button.'

Once a scheme was approved, colour would be applied by the JTAs to their original scale drawings, which they would then deliver by bicycle to a despatch centre outside Leamington, whence they would be sent to the Admiralty for approval before being allocated to the relevant ship. Whenever a ship was dockside for repairs, the camouflage team was instructed by the Admiralty to base its analysis on the ship's next posting. A design was then tested and approved, and a scheme duly executed. The newly camouflaged ships would

142

subsequently be inspected at sea and from the air by Lieutenant-Commander Yunge-Bateman and his deputy. According to Felicity Fisher, no scheme was ever altered or redesigned.

It is clear that the official Admiralty colour range for HM ships was developed by Robert Goodden. It consisted of a range of nine meticulously-researched standard greys and blues, and two 'Special colours –

Western Approaches', Peter Scott Blue and Mountbatten Pink, so-called because Admiral Lord Louis Mountbatten had noticed that a Union Castle liner in a curious shade of pink had seemed to disappear more quickly at dawn and dusk (prime U-boat attack time) than adjacent battleships in Home Fleet Grey. This phenomenon was agreed to be true, not on account of colour but of tone. Felicity Fisher recalls that the pink 'went a mid, warm

143

Opposite: Victorine Foot, Camouflaging a Cruiser in Dock, purchased in 1943. Foot had her own studio space at Leamington, and was probably the JTA who applied this particular camouflage scheme to the original model of HMS Newcastle.
Imperial War Museum

Right: Ratings at Work Camouflaging the Funnel of a Battleship, 1941.
Imperial War Museum

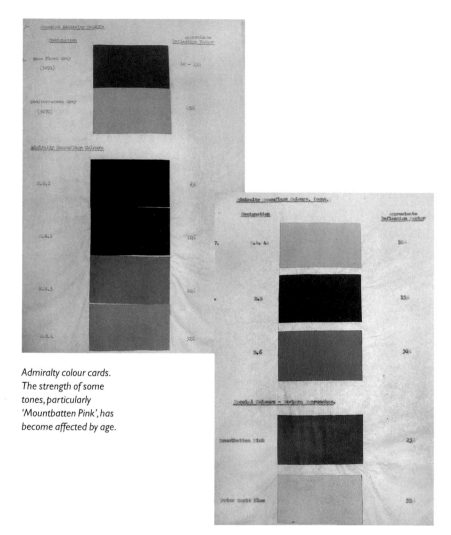

*Admiralty colour cards.
The strength of some
tones, particularly
'Mountbatten Pink', has
become affected by age.*

grey when seen through a telescope'. A 'top secret' note
from the Director of Staff Duties and Training Division,
Naval Staff, dated 17 September 1942, concludes that
after experiments had been carried out at Leamington,
there was 'nothing to choose between pink and an
equivalent tone of grey'. Identical models were tested,
one in a pale pink, blue and green scheme, another in
neutral grey on the starboard side, and 'light Admiralty
Design 23/4' on the port side. It was found that there
was no difference in visibility between the models in
most lighting conditions, although the Admiralty design
worked better downlight in sun and moonlight.[14] A note
about trials at Scapa Flow suggests, conversely, that the
pink could be disadvantageous.[15] However, Mountbatten
Pink appears to have been included in the Admiralty
range, perhaps out of deference for the Admiral!

One of the technical assistants' regular tasks, as
part of the research contributing to the design of
camouflage schemes, was to 'measure the colour of the
sky' every day, using an ingenious instrument made
entirely of painted card. It consisted of a strip with two
rectangular openings, one through which the sky was
viewed, and the other beside it revealing a section of
another long painted strip graded in colours from grey
to blue, which was matched with the current appearance of the sky. This rather unscientific-sounding
device was created by Camouflage Officer 'Bill' Schuil,
an optometrist who had been brought into the team
from a position at the General Electric Company as a
result of a disagreement between Oliver Grahame Hall
and the Deputy Director of Camouflage at Leamington.

Training & Staff Duties Division,
Naval Staff.

TSD 6955/43.

21st June, 1943.

Dear Dr. Ditchburn,

When I visited you at A.R.L. recently you
suggested that it might be worthwhile to measure the
amount of Purkinje effect in some sample blue pigments,
and you asked me to obtain some samples for you. I
enclose three patterns of fairly strong blues labelled
on the back as follows:-

    A. Cobalt Blue (Oil colour)
    B. "Cobalt Tint" (Poster colour)
    C. Ultramarine Blue (Pinchin Johnson's
       Matt Camouflage paint).

These are the only blues I have been able to obtain so
far and as my enquiries for others do not seem to be
progressing very well I think it is best to send you
these three and see what the result of your measurements
of them is.

The Cobalt Blue (A) is artists' oil colour
from a reputable artists' colourman mixed with what
artists call Flake White, which I believe is a white
lead pigment. Presumably it is a genuine Cobalt Blue.
It is ground in Linseed Oil and Turpentine and, I am
afraid, very unevenly painted on by myself, which
makes it rather difficult to avoid specular reflections.
I hope you will be able to get over this difficulty.
The next one, "Cobalt Tint" (B) is an ordinary common
Poster colour supplied by Winsor & Newton's under that
name. This would not be a very high quality pigment
and there is no guarantee that a colour so described
is really manufactured from the pigment whose name is
given to it. However, it is a good vivid blue; in fact

/it is...

it is rather more vivid than the genuine Cobalt,
although the fact that one is ground in oil, and the
other not, makes comparison difficult. The third Blue,
Pinchin Johnson's Ultramarine is a Blue which is at
present being used in camouflage designs and has a
nominal reflection factor of 10%; the other two are
somewhat lighter.

I shall be interested to hear the result of
your measurements and hope that these samples will
prove suitable for the job.

Yours sincerely,

Dr. Ditchburn,
    Admiralty Research Laboratory,
      Teddington,
        Middlesex.

A letter from Robert Goodden dated June 1943 to the Admiralty Research laboratory regarding experiments with blue pigment samples.

The latter could not understand what a bunch of artists, 'shirkers and thoroughly incompetent', were doing there. A furious Hall stood up to him and told him that if he disliked what they were doing, the artists would far rather be at sea 'than sitting on their bottoms in the Admiralty'. In his view, if the official could not accept the judgement of practitioners who had spent their lives making and recording careful observations, perhaps a scientist should be attached to the Design Section. This justified outburst resulted in the appoint-ment of Schuil, a clever young scientist who soon became a popular member of the team.

Schuil's precise use of telephotometry (the meas-urement of light) was to prove, in fact, the accuracy of the artists' approximations of particular tonal values, which often turned out to exactly match his scientific measurements. He was instrumental in setting up the optical equipment used with the tank to assess camou-flage needs. One of the general conclusions drawn as a result of constant research was that ships, if painted in

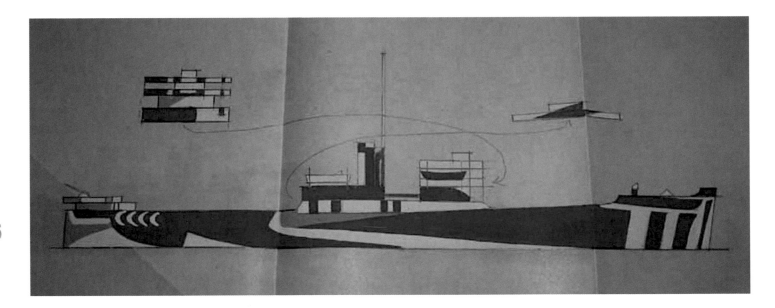

146

a lighter colour other than the customary grey, would be much less conspicuous. Merchant ships were the vessels in discussion, but questions were also raised in Parliament about the Peter Scott 'Western Approaches' scheme for fighting ships – if white was the colour most visible in a blackout, how could the opposite work at sea? Schuil, with his skills in telephotometry, was able to prove that a white ship would most closely correspond with the sky behind it, thus camouflaging it effectively, and that darker objects in colours such as Home Fleet Grey showed up much more prominently.

Victorine Foot records that Peter Scott regularly submitted unsolicited designs intended for vessels in Antarctic waters, and although these often initially met with opposition they were frequently put into use, perhaps proving the advantage of employing a camoufleur endowed with a knowledge of natural history.[16]

Science played a prominent part in an experiment which began with one of Robert Goodden's ideas. Goodden was 'an architect with a scientific background, quiet-spoken, humorous and cultured'.[17] Although he did not rule out the effectiveness of disruptive schemes, 'what could be done with some success was to reduce the range at which a distant ship could first be sighted by painting it to an overall tone to suit the atmospheric conditions in which it would be most valuable for the ship to escape notice'. He found this study fascinating, although success was obviously never more than marginal except in favourable conditions 'like dark overcast night and a hazy horizon by day'.[18] At that time, Goodden had developed an interest in the physiology of vision and had read about a theory known as the 'Purkinje Effect', which causes the peak sensitivity of the human eye to shift towards the blue end of the spectrum at low illumination levels. As a result of his interest in this, he wrote a paper for the

Thomas Rathmell: Design for Inclination, *showing a battleship boldly camouflaged in intense bright blue.*

Director of Scientific Research suggesting that 'as the correct tone of paint for one set of lighting conditions was the wrongest possible in another set of conditions' this could be mitigated to some extent by painting ships 'the strongest, purest blue obtainable!'

To general surprise within the Admiralty, the scientists agreed that this was an idea worth pursuing. Each of the Navy's several Commands was invited to nominate two ships or vessels to be painted brilliant blue. Signals came back nominating two of each Command's least important vessels for the experiment – excepting the Commander-in-Chief, Home Fleet, who nominated a heavy cruiser, HMS *Berwick* (sister ship to HMS *Belfast*), and a destroyer. Goodden spent the next month or so in 'a state of absolute terror' as the newly painted ships went about their duties with the prerequisite that they should report any experiences which could be 'attributable to their unorthodox colour. Mercifully one of the few reports that did come in was from HMS *Berwick*, telling us that she had expected to be sighted and attacked but apparently had not been seen, and she thought the escape might have been due to the way she was painted.' Guy Hartcup, however, points out that the test may not have been properly carried out as *Berwick* was painted in a disruptive pattern and not a uniform blue colour as ordained.

The idea of disruptive camouflage in light, intermediate and dark configurations, continued to remain in favour with Lieutenant-Commander Yunge-Bateman at Leamington, but other methods were constantly being considered. The idea of countershading, inspired

147

Victorine Foot: Robert Goodden, Directorate of Camouflage (Naval Section), Leamington Spa, 1941.
Imperial War Museum

148

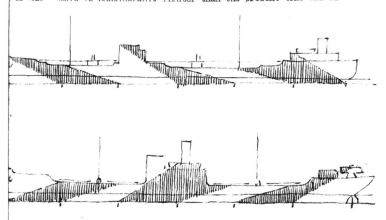

T.S. .1507/43.   Painting of Merchant Ship   *Enclosure B to T.S.P*   Proposal to lighten tone. 15·13/43

Much consideration has been given to the present painting of Merchant Ships in view of the increasing knowledge of the use of paint for concealment. A report RE/Cam/N.5 from D.S.R's representative attached to the R. & E. Department of the Ministry of Home Security in S.R.E.1557/42 deals with this subject.

2.   D.S.R's remarks at A in para.3 of his minute dated 14th September on S.R.E.1557/42 indicated that the best tone for camouflage of merchant ships for concealment is a reflection factor of 25%. This is considerably lighter than the present tone and is

Drop plumb lines over the ship's side so as to divide the light waterline into six equal parts. These lines should be dropped from the highest part of the ship at their station, disregarding the funnels.

2.   Mark on the side and superstructure the slanting lines which join lower end of each plumb line to the top of the next.

3.   Paint the spaces between the slanting lines alternately light and grey as shown. The space at the bow and stern must be light, as in the diagram.

4.   The slanting lines may be drawn either way, i.e. sloping downwards towards the bow or downwards towards the stern.

5.   The Funnel and masts must always be light grey.

6.   The paints to be used are:-

A simple method for camouflaging battleships involved painting the superstructure in equally-proportioned slanting sections in light and dark grey.

by animal camouflage in nature, was gaining popularity. This again used white as a key colour and was approached in two ways. The simpler use of the principle was to paint the lower underwater regions of each vessel white, in the same way that an animal's underbelly is paler than its upper parts. However, even if a ship was able to merge tonally with its background, the shadows caused by its structural planes would be an immediate giveaway to an observer, and this is where 'peripheral' countershading played a part: areas often in deep shadow would be painted white, to flatten their three-dimensional appearance. In the words of a paper written by Yunge-Bateman at the end of 1942, 'Countershading consists in painting the object in such a way as to reverse the normal disposition of light and dark, so that under normal lighting shadow and highlight are cancelled out.'[19]

In situations where a ship was clearly visible and impossible to disguise, the only answer was to baffle the enemy. The camouflage team began to develop methods designed to confuse the ship's identity and its inclination (path of travel). The most successful way of doing this was found to be a simple scheme of painting the ship in two halves – one end light, the other darker in tone, and divided by an oblique line. This would mean that at least one end of the vessel would probably be inconspicuous at any given time, and the range of visibility reduced, thus making it more difficult to target. Countershading would bring white into play as a third colour with 'G45', a greenish grey, and 'B15', a bluish grey. This 'uniform' camouflage appealed to the

Navy's tradition that a ship should look smart – 'anything that looks like blots and patches has to be avoided' – and it also lightened the ship's load in terms of the amount of paint carried on board for repairs. Another successful 'dodge', according to Felicity Fisher, was to paint a 'sham dark prow towards the bows, against a paler grey ... so that the enemy might think it was a smaller vessel altogether'.

The Leamington team was responsible not only for HM ships: its task was also to advise on camouflage for merchant vessels. An order passed in 1943 to merchant ships' crews suggested a treatment whereby the craft would be divided into six sections, alternately light and darker in tone, divided by slanting lines in either direction. This method would appear to fall somewhere between disruptive and 'uniform' camouflage design.

In addition to naval and merchant ships, the camouflage department had to deal with instructions for tank and troop landing craft, and other vessels of war. David Pye and R. D. Russell worked on designs for both British-based landing craft (light medium-toned camouflage) and craft based abroad (dark medium tone).[20] Designs were created solely for concealment, as the issues surrounding confusion and inclination were irrelevant in this case. Colour was the most important element – craft would need to be hidden close to shore, usually at night in natural conditions.

The manufacture of dummy craft for decoy operations (in particular those leading up to D-Day) would be the Army's task, being included in preparation work for large deception plans, but operated by local naval staff. Protection of 'shore establishments' – dockyards, naval airfields, stores, armament and oil depots, and so on, was the responsibility of the Admiralty in conjunction with the Ministry of Home Security.

However, naval 'starfish' lighting sites, such as the two at Falmouth and numerous other coastal targets, became the direct responsibility of the Admiralty. Depending on the size of the installation, shifts of petty officers would man these sites – at least two on duty day and night, taking responsibility for the maintenance, protection and operation of the decoy lighting which was generated by diesel engines. Naval 'starfish' functioned in a similar manner to those used to such superb effect by the RAF. In this case, they had two functions. Firstly, to replicate permitted illumination (dockyards, street lights) and therefore draw bombs away from the actual targets and secondly, to fake 'leaky' lighting as in a sloppily-executed blackout situation. Owing to his extensive experience of creating dummy airfields, Colonel John Turner was again brought in to advise on this particular area of naval camouflage.

Evidently the decoys, particularly those at the naval bases at Portsmouth and Plymouth, were very successful. The day after a raid on Plymouth fires were lit in the actual dockyard to deceive enemy reconnaissance reporters into thinking the attack on the decoy had worked. Bomb craters on dummy sites were covered with netting until it was possible to repair them. An Air Ministry paper of the time records that all naval staff operating these installations were confident and helpful.[21]

150

Camouflage for the Fleet was not restricted to northern waters. In November 1943 David Pye began a series of experiments for camouflaging ships in a possible tropical scenario, in the event of war continuing in the Far East. In this case, he designed canvas 'umbrella' structures which could be easily transported and assembled. These would obscure the silhouette of a boat in normal seagoing camouflage moored alongside land, waiting to spring an ambush or simply avoiding attack. The structures were made of canvas stretched on a frame of light rigid battens, and painted in tones of buff, green and black to be compatible with a very different landscape. Pye's concerns were also for the efficiency of Leamington designs in tropical light conditions, and it was agreed that Schuil's scientific methods should be tested not only at home but also in foreign waters. Oliver Grahame Hall was therefore instructed to undertake a reconnaissance tour in order to 'observe meteorological conditions and levels of light illumination' and 'arrange trials ... to advise on the best camouflage for the prevailing conditions'. He and Schuil were sent on a cruiser to Freetown, West Africa, and thence to Lagos, Khartoum, Cairo and Alexandria where they carried out similar experimental work to that generated at Leamington for home waters.

Having been responsible for the camouflage of British coastal forces, R. D. Russell was later sent abroad to oversee measures taken to protect the Navy. In 1943 he was posted to Ceylon to head up the Admiralty camouflage team there, in order to apply the scientific

Ruskin Spear: David Pye, 1974. Oil on board.
Royal College of Art Collection

and design theories developed at Leamington to the local atmospheric conditions.

In 1943 the Admiralty issued a confidential reference book entitled *Camouflage of Ships at Sea*, a revised version of a report prepared in 1941 for the Admiralty's Director of Staff and Training Duties. According to Oliver Grahame Hall the book was compiled by Gooden and 'almost approached a literary work'. It was a summary of all naval camouflage methods to date, and put known processes and schemes into a standard framework. This publication was superseded in 1945, by which time camouflage practices had altered as a result of a growing knowledge and use of radar. However, the new paper stated that in spite of radar, 'the primary aim … should be to reduce sighting range to a minimum; confusion of inclination, and/or increasing the difficulty of recognising the type or class of ship were subsidiary objectives'.

The use of the 'Peter Scott' and other Admiralty disruptive schemes was continued during 1942 and 1943 as the sighting range did not appear to have been increased. The critical naval activities of the war took place in the North Atlantic, and by the end of 1943 the Allies had more or less won the 'war of the Atlantic', fighting the enemy away from their crucial supply lines. By the middle of 1944 sufficient evidence had been gathered to show that ships painted with light colour schemes were less easily sighted 'in most conditions of lighting and visibility day and night throughout the world'. Disruptive painting 'failed in its primary object of reducing the sighting range of ships'. Countershading 'had been found useful in warships'.[22]

As a result of the far-reaching research carried out into this area of camouflage, all British war vessels except submarines, coastal craft (auxiliary motor gun, torpedo and anti-submarine boats) and landing craft (which continued to use simple disruptive schemes) were painted with standard non-disruptive camouflage schemes using only two colours each. To this day, the Royal Navy continues to use many of the theories developed in Bath and Leamington by the Admiralty camouflage team.

151

CHAPTER TEN

Conclusion

152

'Together they have made a contribution to the war which has probably proved as useful as it has certainly been fascinating.'[1]

THE PLANNING OF Operation Overlord meant that the camouflage unit in Farnham was given orders to disband by 20 March 1944, with the Training Wing actually closing in May after the 21 Army Group's departure to France. Following the Normandy landings, senior army camouflage officers received a letter from Colonel Buckley, dated 19 August 1944, suggesting that instead of a happy return to civilian life, officers should expect to be asked to continue their military duties in one of three or four ways – possibly in the Far East, possibly at home, to support the 'International Situation'.[2] Points 9, 10 and 11 are succinct:

'Very probably there will be a lot of moaning. It is as well, I think, to grasp this thistle and recognise that the end of the war with Germany may not be the end of the Army for us. If you have any views or preferences about yourself … you may like to let me know. But if you have no particular wishes this letter does not call for an answer.'

A subsequent personal letter from Buckley to Ashley Havinden dated 30 October indicates, however, that experienced camouflage officers were having great difficulty in finding new positions – camouflage operations in the UK were only working on a much-diminished scale; there would be few appointments, and these would be mainly for more recently trained staff. Further correspondence (dated 8 December) suggests that Havinden was having problems in severing his links with the Pool, and Buckley, while trying his hardest to be constructive, tactfully points out 'you are a little old perhaps for India' but that a Middle East posting might be possible. Havinden's reply to these communications shows his understanding of the situation and his appreciation of Buckley's solicitude for his officers' future: 'I am sorry in some ways that I can no longer count myself amongst the exclusive band of happy and ubiquitous camoufleurs!' It must have been very difficult for this special and creative group to find postings appropriate for them to take on until demobilisation.

A decision was made in September 1944 by the War Office that future camouflage appointments should be

Opposite: Anne Newland, watercolour sketch of the rink at Leamington, c. 1942, squared up for enlargement. RAF Museum

Staff photograph taken on the closure of the Leamington headquarters,
September 1944. Second row, seated from left to right: L. J. Watson,
James Yunge-Bateman, Gilbert Solomon, Captain Lancelot Glasson, Henry
Hoyland, Christopher Ironside, 'Johnnie' Walker.
Property of Virginia Ironside

held by staff officers (quite possibly those with some artistic experience) who had been specially trained and not by practitioners who, although expert, had arrived from a non-military background.[3] This decision would suggest that the maverick level of creativity which had been offered by the teams at Farnham and Leamington would no longer be necessary in future theatres of war, and may have been true to a certain extent in situations where technology and science would be needed more than art. Concealment and observation should be instructed 'to all ranks at all stages' as they were in the German Army, as a regular component of a military training, and not as part of a special course.

The system whereby each Service had its own camouflage department, which liaised closely with the works division, was seen to work very well, and in this way each could learn from the others. It was found that in naval camouflage the practice of designing schemes specific to expected attack conditions, however difficult they were to predict, was a possible way forward for static camouflage. Development of existing methods would continue; research into new ideas was essential. Although Guy Hartcup remarks that, by the end of the war, the only patterned item of army equipment was the paratrooper's smock, new research showed that small areas of bright colour on a black background would combine visually into an even tone – a natural evolution for scrimmed netting and similar methods.[4] Anyone familiar with the history of painting would point out that this was the first principle of

List of watercolours submitted by camouflage artists to the Ministry of Information in June 1943, showing pencil notes on prices paid.
Imperial War Museum, Second World War Artists' Archive. Ministry of Home Security scheme for artists in the Directorate of Camouflage

pointillism, thus suggesting that an artist's knowledge might well be invaluable within a camouflage team.

It is certainly true that camouflage in the Second World War could not have developed without the imagination and skill of the artists, architects and designers, as well as the art directors and prop-makers. Julian Trevelyan comments that there were two colliding schools of thought in the execution of camouflage schemes – the architects and artists, who 'favoured the use of real materials for construction, such as tile, brick, stone and timber', and the art directors who created illusions with 'plaster-cast sheets of these same materials'.[5] Hugh Casson pointed out after the war that it was possible sometimes to identify the designer of a particular scheme by the style in which it was completed – but did this defeat the object of concealment and disguise?

The Artists' International Association established in 1934 as a social and political programme for the support of artists in Britain, had since 1938 encouraged the formation of regional subgroups. It was found that Leamington's Artists' and Designers' Collective was easily the most interesting of these offshoots, an occurrence obviously due to the fact that 'between a hundred and fifty and two hundred' artists and designers were working in this small Midlands spa town at the peak period for camouflage – at the time the largest concentration of artistic talent in Britain.[6] While the work offered to the camouflage group tested its members' creativity and usually relied on their special skills as artists and designers, it was important that they were

Opposite: G. Watson, View Across Roofs, Rotol Ltd, 1943. A detailed recording of factory camouflage.
Imperial War Museum

Below: Letter from Wing Commander Cave-Brown-Cave referring to the above, dated 22 June 1943.
Imperial War Museum, Second World War Artists' Archive. Ministry of Home Security scheme for artists in the Directorate of Camouflage

MINISTRY OF HOME SECURITY
DIRECTORATE OF CAMOUFLAGE
REGENT HOTEL, LEAMINGTON SPA

'.: Leamington Spa 2190

Your Ref
Our Ref

22nd June, 1943.

Dear Mrs. Murdoch,

I have discussed with the senior officers of the Design Section what subjects ought to be painted in order to record the special items of camouflage technique. We think that five examples would be necessary but sufficient for this purpose:-

1. Dollis Hill. A building of special importance in a built-up area.
2. Yeadon Aircraft Factory. A specially built aircraft factory of the "totally concealed" type. *or Rotol, a specially conceals normal factory.*
3. Brooklands, where a remarkable scheme was carried out at very short notice to conceal the aircraft factories, and especially the racing track.
4. Aircraft on "patch hides" intended to conceal aircraft standing on an aerodrome.
5. A group of ships at sea, showing the effectiveness of present methods of camouflage.

There are nine of the original volunteers who have not yet been released for painting. It is, however, important that these special subjects should be treated as satisfactorily as possible, and I therefore propose to let it be generally known that your Committee is anxious to have examples of these five subjects, and would be comparatively unlikely to take pictures of different subjects unless they were of exceptional artistic merit.

I would, so far as possible, give leave to those who have asked for it, but have not yet been released. On the other hand, several of our most experienced Camouflage

/Officers

Mrs. M.W. Murdoch,
Ministry of Home Security,
Whitehall,
LONDON, S.W. 1.

31

able to have time off in quieter moments to recharge these talents. The Ministry of Home Security recognised this (although some were of the opinion that all energies should be devoted to the war effort) and in 1942 approved a 'scheme for artists in the Directorate of Camouflage' whereby camouflage staff were allowed a month's paid leave to paint a picture which would record representative work being done by the Directorate. A letter from Robin Darwin to the Press and Public Relations branch of the Ministry suggests that artists would be selected on merit and 'probable promise of success'.[7] All paintings submitted went through a censorship process, in case they should be 'detrimental to the national interest' (David Pye acted as the censor for Admiralty subjects), and the War Artists' Advisory Committee would then be able to select a group of paintings which would be a record of the camouflage work done during the war. Artists would be paid for these pictures only if their work was chosen; two batches were submitted in 1943 and a further one in early 1944. Many of these paintings are still in existence, a substantial number being housed at the Imperial War Museum; they act as an enlightening back-up to the work of the official war artists, and they do indeed illustrate the inspiration and diversity behind the numerous camouflage schemes.

Interestingly, there were two camouflage-related subjects that the War Artists' Advisory Committee recommended to the war artist Eric Ravilious, which were never painted. The first was a proposal that he should record an experiment where the Fire Service was to spray chalk railway cuttings with ink to disguise them – to the probable detriment of the Brigade's hoses and equipment! However, the Committee decided that this experience was not worth recording – and in any case Ravilious would have been unhappy working purely from photographs. The second came from a suggestion made by the artist himself on the strength of his pre-war series of paintings of chalk hill figures, that he should record the camouflage operations for these recognisable and therefore vulnerable landmarks. Although Ravilious found the idea attractive, he could not execute it as there was no documentation recording the work carried out and, contrary to the expectations of the Committee, he was unable to work from a mere description of how it was done.

Another scheme whereby the artistic skills of the camouflage designer were put to good use was in the series of murals created for government-run public organisations such as British Restaurants and War Canteens set up in community centres around the country. Felicity Fisher recalls that the painters Stephen Bone and Mary Adshead (both working at the Directorate of Camouflage) painted murals for the British Restaurant in Leamington, as did Edwin La Dell, while she and Victorine Foot decorated the bar of the Regent Hotel ('for no money!'). Over 60 art schools became involved nationwide in the production of the murals, supported by the Ministry of Food and local authorities. Subjects were scenes from wartime life such as 'Carrying on with Domestic Duties after an Air Raid', 'Scene on a Trainer Aerodrome' or 'Apple Orchard'. Julian Trevelyan painted a series of nautical-related schemes for the British Restaurant in Hammersmith.[8] Like other artists, he also managed to exhibit work achieved while stationed in the Middle East and North Africa (where he spent 'lots of time hanging about, drawing and painting') in a high-profile Bond Street gallery.[9]

These morale-boosting public art schemes sponsored by the government were an essential part of the war effort. Whether the country was in immediate danger or not, it was important that people continued to feel optimistic and confident. Furthermore, part of the purpose of civil camouflage was to deal with this issue, and in fact it was assumed 'by the Air Staff' that a great proportion of enemy camouflage was designed with this in mind.[10] Robin Darwin alluded to the

158

consequence of camouflage on the morale of factory workers – 'there can be no doubt of its value and tonic effect in times of heavy raiding'.[11]

Did the original and sometimes scarcely credible work of the camouflage artists and designers have the planned effect? To what degree were they responsible for victory? In Darwin's words, 'the most serious disability suffered by those who practice or are responsible for Camouflage … is that, with rare exceptions, they can never know whether and how often their schemes have proved operationally successful'.[12] Sometimes it was, of course, distressingly easy to see where the camoufleurs had failed. The results of the 'sister art of decoy', drawing attack away from the target, were simpler to measure, and there is no doubt that the success of sensational deceptions such as the battle for El Alamein and the Normandy landings, was due, in addition to the strength and courage of the fighting forces, to the combination of fine-tuned military strategy and lively artistic imagination. As a pertinent example of this, Ashley Havinden, in a discussion with a German officer after the war, learned that the 'Pluto' operation had been totally successful in that the enemy had been entirely oblivious to it.

On the disbanding of the Directorate of Camouflage towards the end of the war, Wing Commander Cave-Brown-Cave gave Stephen Bone the task of recording camouflage experiences in 'policy, design and methods of execution' for a first-hand written summary of camouflage practice. This information was to be included in a textbook and treatise on the camouflage

experience. The result, *An Outline for Organisation of Static Camouflage in Future War*, was eventually compiled and written by Lieutenant-Colonel Francis Wyatt, who promptly scrapped all of Bone's groundwork in favour of a dry and military approach, complete with a number of inaccuracies and omissions, particularly with reference to design. Robin Darwin's comments on the book, written in a letter to Cave-Brown-Cave, list a series of headings tellingly omitted by Wyatt, including 'Viewing Room' (a facility unique to Leamington), 'Models' of schemes and the arguments for and against them, 'Netting' (never properly evaluated in the UK), and not least 'Design'. As Darwin says, 'perhaps the most important of all and probably the hardest … a careful appreciation of the changes and developments in design since 1939, which are manifest, would be stimulating and revealing'.[13] Darwin remained loyal to his camouflage compatriots to the end, always recognising the value of their ideas and achievements. Cave-Brown-Cave's own comments are disparaging: 'the writers have not clearly visualised a human reader, with the result that the 'design' section, for instance, has an almost bizarre unreality. I am not advocating a purpler prose (though … the present prose is, like the night landscape, of 'a uniform greyish tone with blurs here and there').'[14]

In spite of the achievements of the Leamington camoufleurs, it seems that the military establishment still found it hard to acknowledge the extraordinary significance of their productivity. Christopher Ironside sums up Wyatt's failure to include an adequate account

160

CONSTRUCTION

Illustration from
Concealment of New
Buildings.
HM Stationery Office

of camouflage design activities in a letter to Cave-Brown-Cave written on 5 September 1944.[15] 'Design has been an important subject with us and it is most desirable that the accumulated experience of designers should be recorded in a coherent statement ... very little is said about design ... in 4 (General notes on design) there is almost nothing about the subject.'

Ironside fails to avoid a little cynicism in his reference to Wyatt's fleeting summary of the 'Principles of Pattern', which occurs 'alongside specifications for Hessian and steel wool netting, the darkening of concrete and other interesting bits of information including a technical note on Water Coverage'. However, the most glaring omission in Ironside's view

Exhibit demonstrating the use of rubber treatments in camouflage from the 'War to Peace' section in the 'Britain Can Make It' exhibition, 1946.
Property of David Medd

is the lack of illustrations or diagrams: 'Whatever the reasons it is difficult to understand how anyone could imagine that a report on such a 'visual' subject as camouflage could be adequate without pictures – either photographs or drawings … only one in the main body of the book and this is not important.'

It is impossible to judge accurately the results of the work carried out by the Directorate, particularly during the first part of the war when everything was at a purely experimental stage, and the camouflage team had to learn fast. But when the ideas of the camoufleurs were executed precisely to plan (often allowing for a little harmless artistic licence!) the assaulting side would be enormously disadvantaged, sometimes fatally.

As the war developed, it became clear that deception was becoming more crucial than disguise, and combat began to include the game of misinformation. Psychological trickery played as great a part as visual illusion, and the two were used together to astounding effect.

War teaches many lessons, some of which should always have an effect on future thinking. Hugh Casson's fear was that camouflage would be 'a certain war casualty' particularly in relation to architecture: 'few industrialists, it can safely be presumed, prefer their factories dressed in multi-coloured modesty vests'.[16] But, with hindsight, would it not have been an idea to ensure that buildings were always designed to be in harmony with their surroundings – while challenging modern concepts of design, to retain an organic and sympathetic character? As David Medd points out, the format used by the Camouflage Training and Development Centre was very strong in its understanding of future requirements: a clear brief, the involvement of industrial links at an early stage, and the use of experts in design and manufacturing. After the war this concept was taken forward into Local Authority architecture, particularly for schools.[17] Many techniques developed in the interests of camouflage were taken up by industry once hostilities had ceased.[18] Lasting business ventures

and friendships, too, were formed in this environment. Not least, many of the Leamington and Farnham camoufleurs went on to teach at the Royal College of Art under its enlightened new principal Robin Darwin. The list of senior staff in the early years of the College's reincarnation includes Hugh Casson, Robert Goodden, Richard Guyatt, Janey Ironside (wife of Christopher), Edwin La Dell, David Pye, Brian Robb and R. D. Russell. Numerous members of the College's large pool of visiting art and design staff had also held prominent camouflage posts. Many of these were to become key figures in the important morale-boosting post-war exhibitions, 'Britain Can Make It' and the South Bank utopia of the Festival of Britain.

Whatever the merits, strengths and weaknesses of the genre, camouflage was a unique feature of the war in that it brought together an extraordinary group of like-minded individuals who combined passionate eccentricity with practical knowledge. In the words of Hugh Casson, 'with grace and gaiety' it performed a more than useful war service. Although not magic, it nevertheless united trickery and shrewdness with skill, elegance and often humour to create illusions which were astonishing in their audacity. The ingenuity of these concepts played a crucial part in assisting the Allied forces to win the Second World War.

Edward Ardizzone, The Senior Common Room of the Royal College of Art, *1951*.
Royal College of Art Collection

Notes

Abbreviations

ADM   Admiralty
AIR   Air Ministry
DEFE   Defence
HO   Home Office
IWM   Imperial War Museum
JOT   Julian Otto Trevelyan
MOD   Ministry of Defence
PREM   Premier
TNA   The National Archives
WO   War Office

Chapter One

1. Julian Trevelyan, 'The Technique of Camouflage', *The Architectural Review*, September 1944.
2. Roy R. Behrens, *False Colors: Art, Design and Modern Camouflage* (Iowa, Bobolink Books, 2002), p. 171.
3. Behrens, *False Colors*, pp. 38–54.
4. Colin Dobinson, *Fields of Deception* (London, Methuen, 2000), pp. 2–3.
5. Dobinson, *Fields of Deception*, p. 3.
6. TNA/PREM 4/97/3.
7. Behrens, *False Colors*, p. 89.
8. Behrens, *False Colors*, pp. 105–106.
9. TNA/PREM 4/97/3.

10. C. H. R. Chesney, *The Art of Camouflage* (London, Robert Hale, 1941), p. 13.
11. Robin Darwin, 'The Role of Artists in Camouflage', 18 February 1943. TNA/HO 186/1648.

Chapter Two

1. 'The Camouflage of Vital Factories and Key Points'. IWM, Ministry of Home Security, 83.9(41)3/5.
2. 'The Camouflage of Vital Factories and Key Points'. IWM 83.9(41)3/5.
3. TNA/WO 227/48.
4. Mass-Observation Archive, University of Sussex, Art/2/B.
5. Mass-Observation Archive, Art/2/B.
6. Guy Hartcup in conversation, 3 December 2003. Hartcup is the author of *Camouflage: A History of Concealment and Deception in War* (Newton Abbott, David & Charles, 1979).
7. TNA/AIR 2/3/94.
8. IWM 83.9(41)3/5.
9. From Christopher Ironside's obituary by Fiona MacCarthy, the *Guardian*, 18 July 1992.
10. TNA/WO 199/1630.

11. Ursula Mommens, Robin Darwin's sister and Julian Trevelyan's first wife, in conversation with author, April 2003.
12. Ursula Mommens.
13. Julian Trevelyan, 'The Technique of Camouflage', *The Architectural Review*, September 1944.

Chapter Three

1. James Gardner, *The ARTful Designer: Ideas Off the Drawing Board* (London, Centurion Press, 1993), p. 97.
2. Julian Trevelyan, IWM Sound Archive, 1987.
3. 'Painting the Picture'. R. H. Kelly in a lecture for the Camouflage Development and Training Centre, 1944. IWM 74/165/1 Havinden.
4. Julian Trevelyan, 'The Technique of Camouflage', *The Architectural Review*, September 1944.
5. IWM 74/165/7 Havinden.
6. Geoffrey Barkas, *The Camouflage Story* (London, Cassell, 1952).
7. John Lewis, *Such Things Happen* (Stowmarket, Unicorn Press, 1994), p. 56.
8. National Sound Archive, F 6755 (side A).

9. Roland Penrose, *Scrapbook: 1900–1981* (London, Thames & Hudson, 1981), p. 131.
10. Antony Penrose, *Roland Penrose: The Friendly Surrealist* (London, Prestel, 2001), p. 102.
11. TNA/HO 186/975.
12. Lewis, *Such Things Happen*, p. 60.
13. David Medd to author, November 2003.
14. IWM 74/165/1 Havinden.
15. Henry Buckton, *Artists and Authors at War* (Barnsley, Pen and Sword, 1999), p. 92.
16. Coastal Command directed the enormous number of small vessels, including torpedo boats, gunboats and anti-submarine boats.
17. IWM Documents, 202 90/6/1 & 1a.
18. Barkas, *The Camouflage Story*, p. 153.
19. IWM 7762 74/165/1 Havinden.

Chapter Four

1. TNA/HO 186/975.
2. 'Artists in Camouflage'. TNA/HO 186/1648.
3. Mass-Observation Archive, University of Sussex.
4. Richard Guyatt to author, July 1998. Christopher Ironside's wife was Janey Ironside, later to become head of the school of fashion at the RCA.
5. Julian Trevelyan, 'The Technique of Camouflage', *The Architectural Review*, September 1944.
6. Hugh Casson, 'Art by Accident', *The Architectural Review*, September 1944.
7. John Lewis, *Such Things Happen* (Stowmarket, Unicorn Press, 1994), p. 59.
8. A paper by Tom Purvis, *Art and Industry*, November 1939, p. 200, quoted by Julian

Trevelyan in 'Impact of War on Art – War Jobs'.
9. TNA/HO 186/975.
10. TNA/HO 196/15.
11. TNA/HO 186/1987.
12. TNA/HO 199/1652.
13. TNA/HO 196/15.
14. TNA/HO 196/15.
15. Casson, *The Architectural Review*, September 1944
16. TNA/HO 217/5.
17. Property of Virginia Ironside.
18. Lt-Col Francis Wyatt, *The Principles and Organisation of Static Camouflage*, 1944. TNA/HO 186/1987.
19. TNA/HO 186/975.

Chapter Five

1. Colin Dobinson, *Fields of Deception* (London, Methuen, 2000), pp. 5–6.
2. 'Decoy and Deception'. Air Ministry, Air Historical Branch RAF Monograph (first draft), 1945.
3. Dobinson, *Fields of Deception*, p. 19.
4. 'Decoy and Deception'. Air Historical Branch.
5. Dobinson, *Fields of Deception*, p. 24.
6. IWM 74/165/1.
7. TNA/HO 186/975.
8. 'The Royal Air Force Builds for War'. MOD, Air Historical Branch, 1997.
9. TNA/HO 186/975.
10. 'The Royal Air Force Builds for War'. Air Historical Branch.
11. José Manser, *Hugh Casson, a Biography* (London, Viking, 2000), p. 90.

12. 'The Royal Air Force builds for War'.
13. Manser, *Hugh Casson, a biography*, p. 99.
14. 'The Royal Air Force Builds for War'.
15. 'The Royal Air Force Builds for War'.
16. Carola Zogolovitch to author, January 2004.

Chapter Six

1. TNA/WO 199/2548.
2. National Sound Archive, F 6755 (side A).
3. TNA/WO 199/2548.
4. Norman Scarfe to author, March 2004.
5. TNA/WO 199/1630. *The Fortnightly DO* was a bulletin for unit camouflage instructors.
6. TNA/WO 199/1630.
7. IWM 74/165/1 Havinden.
8. IWM 74/165/1 Havinden.
9. IWM 74/165/1 Havinden.

Chapter Seven

1. Stephen Sykes, *Deceivers Ever: Memoirs of a Camouflage Officer 1939–1945* (Speldhurst, Spellmount, 1990), p. 16.
2. Geoffrey Barkas, *The Camouflage Story* (London, Cassell, 1952), p. 58.
3. IWM 74/165/1 Havinden.
4. Julian Trevelyan, 'The Technique of Camouflage', *The Architectural Review*, September 1944.
5. John Morton to author, May 2003.
6. William Murray Dixon to author, August 2002.
7. Trinity College, Trevelyan Archive, JOT 58/2.
8. IWM 74/165/1 Havinden.
9. TNA/WO 169/1.
10. TNA/WO 199/1630.

11. Trinity College, Trevelyan Archive, JOT 58/1.
12. Barkas, *The Camouflage Story*, pp.120–121.
13. Barkas, *The Camouflage Story*, p. 145.
14. Trinity College, Trevelyan Archive, JOT 58/1.
15. Sykes, *Deceivers Ever*, p. 54.
16. Trevelyan, 'The Technique of Camouflage', *Architectural Review*.

Chapter Eight

1. 'The War Diary of the Camouflage Service of Headquarters 21 Army Group', May 1944. TNA/WO 171/3832, appendix B.
2. 'War Diary'. TNA/WO 171/3832, appendix B5.
3. 'War Diary'. TNA/WO 171/3832, appendix B3.
4. 'War Diary'. TNA/WO 171/3832, appendix D1.
5. David Sutton, 'Bodyguard of Lies', *Fortean Times*, FT 185, April 2004.
6. TNA/AIR 20/4259.
7. TNA/AIR 20/4259.
8. Seymour Reit, *Masquerade: The Amazing Camouflage Deceptions of World War II*, (London, Hale, 1979), p. 39.
9. James Gardner, *The ARTful Designer: Ideas Off the Drawing Board* (London, Centurion Press, 1993), p. 120.
10. TNA/AIR 20/8153.
11. IWM 74/165/7 Havinden.
12. Reit, *Masquerade* (1999), p. 33-34.
13. 'Letters from D-Day', the *Guardian*, 28 May 2004.
14. Sutton, *Fortean Times*.

Chapter Nine

1. David Williams, *Liners in Battledress: Naval Camouflage 1914-45* (London, Chatham, 2001), p. 73.
2. Williams, *Liners in Battledress: Naval Camouflage 1914–45*, p. 74.
3. TNA/ADM 1/17004.
4. Peter Scott, *The Eye of the Wind* (London, Hodder & Stoughton, 1961), p. 392.
5. TNA/ADM 1/17004.
6. TNA/ADM 204/2537.
7. TNA/ADM 204/2537.
8. TNA/HO 186/975.
9. Martin Muncaster, *The Wind in the Oak* (London, Robin Garton, 1978), p. 60.
10. TNA/ADM 1/17004.
11. TNA/ADM 1/11728.
12. Muncaster, *The Wind in the Oak*, p. 60.
13. Robert Goodden to author, July 1996.
14. TNA/ADM 212/122.
15. TNA/DEFE 2/878.
16. IWM Sound Archive, accession no.13516, 20 October 1993.
17. Muncaster, *The Wind in the Oak*, p. 61.
18. Robert Goodden to Fiona MacCarthy, May 1986.
19. 'Notes on Naval Camouflage'. TNA/ADM 204/2537.
20. TNA/ADM 1/13444.

15. 'War Diary'. TNA/WO 171/3831, appendix E.
16. 'War Diary'. TNA/WO 171/3831, appendix D.
17. 'War Diary'. TNA/WO 171/3831, appendix D.

21. TNA/AIR 20/4259.
22. TNA/ADM 1/17004.

Chapter Ten

1. Robin Darwin, 'The Role of Artists in Camouflage', 18 February 1943. TNA/HO 186/1648.
2. IWM 74/165/7 Havinden.
3. TNA/WO 199/1135.
4. 'The Camouflage of Vital Factories and Key Points'. IWM, Ministry of Home Security, 83.9(41)3/5.
5. Julian Trevelyan, 'The Technique of Camouflage', *The Architectural Review*, September 1944.
6. Lynda Morris and Robert Radford, *AIA: The Story of the Artists' International Association 1933–1953* (Oxford, Museum of Modern Art, 1983), p. 65.
7. TNA/HO 186/1648.
8. Morris and Radford, *AIA: The Story of the Artists' International Association*, p. 65.
9. IWM Sound Archive, accession no. 3172, 1978.
10. TNA/HO 186/1987.
11. TNA/HO 186/1648.
12. TNA/HO 186/1648.
13. TNA/HO 186/1987.
14. TNA/HO 186/1987.
15. TNA/HO 186/1987.
16. Hugh Casson, 'Art by Accident', *The Architectural Review*, September 1944.
17. David Medd to author, November 2003.
18. Dunlop's rubber coatings for hollow floors, for example. David Medd to author, November 2003

Acknowledgements

I would like to thank the following, whose generosity and support have made this book possible:

Dinah Casson, who had the original idea
Professor Sir Christopher Frayling, Professor Wendy Dagworthy, and
    the Research Office at the Royal College of Art
Georgia Dickson, for her exemplary research
Hugh Tempest-Radford and everyone at Unicorn Press
Andrew Dalton and Cartwright Hall Art Gallery, for the loan of the
    beautiful cover image

Admiralty Archive
Ashmolean Museum
Brighton and Hove City Council
British Library
Cassell plc
Churchill Archives Centre, Churchill College, Cambridge
Getty (Hulton Archive)
Government Art Collection
Hyman Kreitman Research Centre, Tate Britain
Imperial War Museum
James Gardner Archive, Design History Research Centre,
    University of Brighton

Julian Trevelyan Archive at Trinity College, Cambridge
Leamington Spa Art Gallery and Museum
National Film and Television Archive
National Sound Archive
QB Archives
Roland Penrose Collection/Lee Miller Archives
RAF Museum
RE Museum
Royal College of Art Collection
Scottish National Gallery of Modern Art Archive
Sir Hugh Casson Ltd
The National Archives

Julia Atkinson, Professor Roy Behrens, Quentin Blake, Peter Dixon, William Murray Dixon, Felicity Fisher, Will Grove-White, Dick Guyatt, Guy Hartcup, Michael Havinden, Lizzie Hutton, Marigold Hutton, Virginia Ironside, Fiona MacCarthy, David Medd, Ursula Mommens, Michael Moody, John Morton, Pat Moss, Elizabeth Mozley, Graeme Raeburn, Chiara Schilska, Richard Slocombe, Juliet Thorp, Emma Tristram, Jenny Wood, Carola Zogolovitch

David Thomas, for his continuing encouragement and IT support.

Select Biographies

Bainbridge-Copnall, Edward
1903–1973
Painter and sculptor

Edward Bainbridge-Copnall was educated at Skinner's College and Collyer's School. He trained at Goldsmiths College and the Royal Academy Schools, and became head of Sir John Cass College (1945–53).

Bainbridge-Copnall was president of the Royal Society of British Sculptors (1961–6).

Barkas, Geoffrey
1896–1979
Filmmaker

Geoffrey Barkas worked for a year in insurance after leaving school.

When the First World War broke out Barkas joined up as an infantryman. After the war he travelled to Canada, where he worked as a farmer's boy, then to California to try his luck as an actor. He became a camera boy instead, advancing to cameraman in 1920, and eventually producer and director. In 1925 Barkas travelled for seven months with the Prince of Wales as official cinematographer, throughout a tour of West and South Africa and South America.

During the Second World War Barkas held the position of director of camouflage, Middle East Forces, Royal Engineers.

Barkas returned to film after the war. He made children's films for the Rank Organisation, including *Dusty Bates* and *The Little Ballerina*. He then joined Shell-Mex and BP in 1946, remaining there as producer until 1956 when he joined Random Film Productions. He was appointed OBE in 1946.

Beddington, Frederick
1896–1979
Stockbroker and art dealer

Frederick Beddington was educated at Wellington College, and studied art at Byam Shaw School during his school holidays.

In 1914 Beddington was commissioned in the Kings Royal Rifle Corps. He was demobilised in 1918 and enrolled for a year at the Slade School of Fine Art. He then spent a year as a bank clerk before becoming a stockbroker.

During the Second World War Beddington worked as a camouflage advisor to the British Expeditionary Forces. After the war he became director of Wildenstein's art dealers, London.

Beddington, John Louis (Jack)
1893–1959
Publicity manager

Jack Beddington, brother of Frederick, was educated at Wellington College and Balliol College, Oxford.

At the outbreak of the First World War he enlisted in the King's Own Yorkshire Light Infantry and served until 1919. He was severely wounded at Ypres.

After the war Beddington worked for the Asiatic Petroleum Company in Shanghai, but was invalided home in 1928. Back in London he was given the job of publicity manager, and took on the same role for Shell-Mex and BP on its formation in 1932.

Since the early 1920s advertising posters had been fixed onto Shell lorries. Beddington made the 'lorrybill' series his own, commissioning fine artists and graphic designers including Graham Sutherland, Paul Nash and Ben Nicholson. He was approached by John Betjeman to publish 13 Shell County Guides, including Betjeman's *Cornwall* (1934) and *Devon* (1936), and John Piper's *Shropshire* (1951).

When the Second World War began he was appointed joint manager of the staff department of the Petroleum Board. In 1940 he became director of the films division at the Ministry of Information, where he remained until 1946. Documentaries produced during this time included *Desert Victory* (1943) and *The True Glory* (1945). Beddington coined the famous anti-gossip slogan, 'Be like Dad: keep Mum'.

After the war Beddington joined the advertising agency Colman Prentis and Varley, where he remained until his death.

Awards include Honorary Fellowships of the Society of Industrial Artists and the Royal Society of Arts. He was also governor of the Foundation for Visual Education, member of the National Advisory Council on Art Education, and member of the council of the Royal College of Art. He was appointed CBE in 1943.

Bone, Stephen
1904–1958
Artist, writer, and broadcaster

Stephen Bone was educated at Bedales School, and was also taught privately by Stanley Spencer, who lodged at the family home. After leaving school Bone travelled with his father before attending the Slade School of Fine Art from 1922. He left, disenchanted, in 1924 and began to illustrate books with wood engravings. In 1928 he was commissioned by the London Underground to paint a mural for Piccadilly Circus station. Bone was made a member of the New English Art Club in 1932.

He joined the Directorate of Camouflage in 1940, and was also an official war artist attached to the Royal Navy.

After the war Bone turned to journalism. He became art critic of the *Manchester Guardian*, and contributed to the *Glasgow Herald* under the pen-name Luggage McLuggage. He participated in BBC Radio's *The Brains Trust* and *The Critics*, as well as the long-running television quiz

programme *Animal, Vegetable or Mineral*. He wrote and illustrated children's books with his wife, Mary Adshead (1904–1995), herself a painter and Slade contemporary.

Publications: *Albion, an Artist's Britain* (1939).

Casson, Sir Hugh Maxwell
1910–1999
Architect

Hugh Casson was educated at Eastbourne College. He trained at St John's College, Cambridge and the Bartlett School of Architecture.

Casson went to work for Christopher Nicholson in 1935, taking over the practice with Neville Conder and Patience Clifford on Nicholson's death in 1948.

He joined the Camouflage Branch, Air Ministry in 1940, remaining there until 1943. In 1944 he moved to the Ministry of Town Planning, rejoining Nicholson in practice in 1946.

He was appointed director of architecture for the Festival of Britain in 1948. This dominated his working life until the opening in 1951. Casson taught at the Royal College of Art as professor of environmental design (1953–75).

He was elected president of the Royal Academy in 1975, made Royal Designer for Industry in 1961 and knighted in 1952.

Clark, Kenneth
1903–1983
Art historian, writer and broadcaster

Kenneth Clark was educated at Winchester College, and won a scholarship to Trinity College, Oxford where he studied modern history.

Between 1925–6 he assisted Bernard Berenson with his work on Florentine drawings. He was asked in 1929 to catalogue Windsor Castle's collection of drawings by Leonardo da Vinci, and in 1931 became director of fine art at the Ashmolean Museum, Oxford.

From 1934–45 he held the position of director of the National Gallery, and during this time was also Surveyor of the King's Pictures. The National

169

Gallery's major works were evacuated to a slate quarry in Wales for the duration of the war.

He was instrumental in setting up the Council for the Encouragement of Music and the Arts (later the Arts Council) in 1939–40, and became its chairman from 1953–60. He was also chairman of the Independent Television Authority (1954–7).

His 1969 TV series and book *Civilisation* made art accessible to a whole generation of the English-speaking public, and in the same year he was made a life peer.

Publications: *The Gothic Revival* (1928), *Landscape into Art* (1949). *The Nude* (1956), *Looking at Pictures* (1960), *Another Part of the Wood* (autobiography, 1974).

Cott, Hugh Bamford

1900–1987
Zoologist and teacher

Hugh Cott was educated at Rugby School, the Royal Military College, Sandhurst, and Selwyn College, Cambridge.

Cott joined the 1st Battalion of the Leicester Regiment and served in Ireland (1919–21). He completed the first part of the Natural Sciences Tripos at Cambridge in 1925.

He made zoological expeditions to south-east Brazil (1923), the lower Amazon (1925–6), the Zambezi (1927), the Canary Islands (1931), Uganda (1952), Zululand (1956) and central Africa (1957).

Cott taught hygiene at Bristol University (1928–32) and zoology at Glasgow University (1932–8). He was Strickland Curator and Lecturer at the University of Cambridge (1938–67) and was lecturer (1945–67) and dean (1966–7) of Selwyn College.

During the Second World War he was a member of the Advisory Committee on Camouflage, and was captain and major with the Royal Engineers, serving in the Western Desert. He was Chief Instructor of the Middle East Camouflage School (1941–3), and GSO 2 (Cam) at the Mountain Warfare Training Centre (1943–4).

Cott was Fellow of the Royal Photographic Society, the Zoological Society and Selwyn College, Cambridge, and a founder member of the Society of Wildlife Artists.

Publications: *Adaptive Colouration in Animals* (1940), *Zoological Photography in Practice* (1956), *Uganda in Black and White* (1959), *Looking at Animals*: *a Zoologist in Africa* (1975), *Adaptive Colouration in Nature* (1940).

Darwin, Sir Robert Vere (Robin)

1910–1974
Painter and teacher

Robin Darwin was educated at Eton College and the Slade School of Fine Art from 1929. He taught art at Watford Grammar School, then at Eton College.

At the outbreak of the Second World War Darwin joined the Directorate of Camouflage, later becoming secretary to the Camouflage Committee.

Towards the end of the war Darwin moved to the Ministry of Town and Country Planning. In 1945 he was appointed education officer of the newly formed Council of Industrial Design.

In 1946 Darwin was made director of the King Edward VII School of Art, Newcastle upon Tyne and professor of fine art at Durham University. In 1948 he became principal of the Royal College of Art, and later became its first rector and vice-provost (1967).

He was elected Associate of the Royal Academy in 1966 and Royal Academician in 1972. He was appointed CBE in 1954, and knighted in 1964.

Felicity Fisher (née Sutton)

1922–2007
Artist

Felicity Fisher was educated at the Convent of the Sacred Heart in Roehampton before training at the Chelsea School of Art (1938–9).

At the age of 18 Fisher joined the Admiralty Camouflage HQ at Leamington as a junior technical assistant, working on camouflage designs for the Fleet. She left Leamington for London in 1944 to work for the Air Ministry, where she wrote a history of air warfare.

After the war, Fisher returned to Chelsea to finish her art training.

170

Foot, Victorine

1920–2000

Artist

Victorine Foot trained at the Central School of Arts and Crafts, the Chelsea School of Art and Edinburgh College of Art.

During the Second World War Foot worked as a junior technical assistant at the CDCE in Leamington Spa. She submitted and sold several artworks to the War Artists Advisory Committee. Foot later married the sculptor Eric Schilsky, who was also based at Leamington.

From 1950–1 Foot taught at Edinburgh College of Art and later at Oxenford Castle School. She held 11 one-woman exhibitions in Scotland and was represented in many group exhibitions.

Gardner, James

1907–1995

Designer and graphic artist

James Gardner trained at Westminster School of Art. He was apprenticed to Cartier in 1923, then employed by the commercial design consultancy Carlton Studios.

Gardner designed advertisements and exhibitions throughout the 1930s, including posters commissioned by Jack Beddington at Shell-Mex and BP.

At the outbreak of the Second World War, Gardner joined the CD&TC, remaining there as an instructor following training. He was also employed as a wartime illustrator by the Ministry of Information.

His post-war work as an exhibition designer included 'Britain Can Make It' (1946) and the Festival of Britain (1951). Gardner also designed a number of museums including the Evolution Museum in Eindhoven (1966) and the Museum of Intolerance in Los Angeles (1989).

He was made Royal Designer for Industry in 1947 and was awarded the medal of the Chartered Society of Designers in 1989.

Autobiography: *Elephants in the Attic* (1983), *The ARTful Designer: Ideas Off the Drawing Board* (1993).

Gardner, William Maving

1914–2001

Designer and calligrapher

William Maving Gardner was brought up in London by his aunt. During his teens he attended lettering classes before embarking on his training at the Royal College of Art in 1935. Here he studied engraving, typography, letter-cutting and stained glass. He designed and engraved the RCA's Silver Medal of Merit, and won a travelling scholarship in 1939.

This trip was cut short by the outbreak of war, and Gardner was posted to the CD&TC in Farnham. He later established the camouflage training school in Edinburgh.

After the war, Gardner was responsible for many of the Royal Mint's coin designs, and also worked in the fields of heraldry, calligraphy and lettering.

Gardner was elected a Fellow of the Royal Society of Arts in 1955 and served on the jury of its industrial design bursary scheme. He became a visiting lecturer at several British art schools, as well as becoming a visiting professor at Colorado State University in 1963.

Publications: *Alphabet at Work* (1982), *William Gardner's Book of Calligraphy* (1988).

Glasson, Lancelot Myles

1894–1959

Painter

Lancelot Myles Glasson was educated at Marlborough College and the Royal Academy Schools.

During the Second World War, Glasson was Chief Camouflage Officer for the Air Ministry and Ministry of Home Security. His role included the recruitment of artists and designers to the Directorate of Camouflage.

Goodden, Robert Yorke

1909–2002

Architect, designer and teacher

Robert Goodden was educated at Harrow School and trained at the

Architectural Association. In addition to architectural commissions, his design output during the 1930s included wallpaper and domestic glass.

During the Second World War Goodden was recruited into the Admiralty's Training and Staff Duties Department, and became Admiralty liaison officer to the Naval Camouflage Unit.

After the war Goodden was at the forefront of the 'homes for people' movement. He joined R. D. Russell in architectural partnership in London. Design commissions included the sports hall of the 'Britain Can Make It' exhibition (1946) and the Lion & Unicorn Pavilion at the Festival of Britain (1951). He designed the silver tea set used by King George VI and Queen Elizabeth at the opening of the Festival of Britain.

Goodden taught at the Royal College of Art as professor of silversmithing and jewellery from 1948, becoming pro-rector in 1967. He was also co-architect, together with Hugh Casson and H. T. Cadbury Brown, of the RCA's post-war building.

He was made Royal Designer for Industry in 1947, Prime Warden of the Goldsmiths Company and Chairman of the Crafts Council. He was appointed CBE in 1965.

Publication (with Philip Popham): *Silversmithing*, OUP 1971.

Gore, Frederick John Pym
1913–
Painter and teacher

Frederick Gore was educated at Lancing College. He trained as a painter at the Ruskin School whilst reading Classics at Trinity College Oxford (1932–4), then at Westminster School of Art and the Slade School of Fine Art (1934–7). Gore spent the summers of 1937 and 1938 painting in Provence and Greece.

At the outbreak of the Second World War Gore joined the CD&TC. After training he was posted to the Eastern Command Camouflage School, Norwich as a camouflage officer.

Gore taught at St Martin's, Chelsea and Epsom Schools of Art (1946–79). He was head of the painting department at St Martins from 1951.

He was elected Associate of the Royal Academy in 1964, Royal Academician in 1973, and chairman of the Royal Academy Exhibitions Committee between 1976 and 1987. He was appointed CBE in 1988.

Publications: *Abstract Art* (1956), *Painting: Some Basic Principles* (1965), and *Piero della Francesca: The Baptism* (1969).

Guyatt, Richard Gerald Talbot
1914–
Teacher and designer

Richard Guyatt was educated at Charterhouse.

Amongst his commissions as a freelance graphic designer in the 1930s were posters for Shell-Mex and British Petroleum (BP).

During the Second World War Guyatt joined the Directorate of Camouflage, where he was involved in the concealment of factories. He later worked on naval camouflage for the Admiralty. Following a reorganisation of the Directorate he became regional camouflage officer for Scotland.

Guyatt was professor of graphic arts at the Royal College of Art from 1948, and became its pro-rector in 1974 and rector in 1978.

He was director of the design company Cockade (1946–8) and consultant designer to Wedgwood (1952–5 and 1967–70). He co-founded the Guyatt/Jenkins Design Group with Nicholas Jenkins in 1972. Guyatt's design work includes commemorative mugs for the Coronation, the Royal Silver Wedding, the Millennium and the Golden Jubilee, stamps for the Silver Jubilee and a crown piece for The Queen Mother's 80th birthday.

Guyatt was awarded the Society of Industrial Artists and Designers' Misha Black Memorial Medal in 2000. He was appointed CBE in 1969.

Autobiography: *Drawn From Life* (1985).

Havinden, Ashley Eldrid
1903–1973
Designer, typographer and painter

Ashley Havinden was educated at Christ's Hospital. He left school with no formal qualifications and took evening classes at the Central School of Arts and Crafts. He trained privately with Henry Moore from 1933.

Havinden entered Crawford's advertising agency as a trainee and remained there for the whole of his career, becoming art director in 1929. He designed the brand image for Simpsons in 1935.

At the outbreak of the Second World War, Havinden enrolled in the Highgate Home Guard Battalion. He also designed posters for the Air Raid Precautions Service and the Ministry of Information. In 1940 Havinden was commissioned as an army camouflage officer and joined the CT&DC. He was promoted to captain in 1943.

Havinden was appointed governor of the London College of Printing in 1950, retaining the role until 1967. In 1964 he was made governor of Chelsea School of Art.

The Monotype Corporation bought his Ashley Script typeface in 1955. He was made Royal Designer for Industry in 1947 and Master from 1967–9, Fellow and later President of the Society of Industrial Artists, and member (1951) and then President (1957–9) of the Alliance Graphique Internationale. He was appointed OBE in 1951.

Publication: *Advertising and the Artist* (1956).

Catalogue: Michael Havinden et al., *Ashley Havinden: Advertising and the Artist*, Scottish National Gallery of Modern Art (2003).

Hayter, Stanley William
1901–1988
Painter and printmaker

Stanley William Hayter was educated at Whitgift Middle School, Croydon, and King's College, London (1918–21).

Hayter worked as a chemist for the Anglo Persian Oil Company in Iran (1922–5). In 1926 he went to Paris, and briefly attended the Académie Julian. In 1927 he established a printmaking workshop, which later became Atelier 17 (so-named in 1933).

Hayter returned to England in 1939 and established the Industrial Camouflage Research Unit. He was unable to take up military service due to injury, and went to New York in 1940, establishing Atelier 17 at the New School for Social Research. The workshop's exhibition at the Museum of Modern Art (1944) brought renown and during Hayter's decade in New York Atelier 17 attracted a number of prominent European artists. Atelier 17 reopened in Paris in 1950.

Hayter was awarded the Légion d'Honneur (1951), made a Chevalier de l'ordre des Arts et Lettres (1968) and Commandeur des Arts et Lettres (1986). He was elected Honorary Royal Academician in 1982, and appointed OBE in 1959 and CBE in 1968.

Publications: *New Ways of Gravure* (1949) and *About Prints* (1962).

Hutton, John
1906–1978
Artist and glass engraver

John Hutton was born in New Zealand and attended Wanganui Collegiate School. Initially Hutton studied law, but at the age of 25 he abandoned this for a career as a self-taught artist. He relocated to England in 1935, where he worked extensively as mural painter.

During the Second World War Hutton served in the Middle East before being transferred, with the rank of major, to work on camouflage strategies in preparation for the D-Day landings.

After the war Hutton continued to paint, and his later work included murals for the Festival of Britain, Buckingham Palace and Orient Line passenger liners. He is especially well-known for his large-scale glass engravings on the Great West Screen of Coventry Cathedral.

Publications: Margaret Brentnall and Marigold Hutton, *John Hutton: Artist and Glass Engraver* (1986).

Ironside, Christopher
1913–1992
Artist, designer and teacher

Christopher Ironside trained at the Central School of Arts and Crafts. He then went to work as an art master at Miss Ironside's School, Kensington, which was run by his aunt and where he himself was educated.

During the Second World War, Ironside was recruited into the Directorate of Camouflage as deputy senior design officer.

After the war he worked for the Ministry of Town Planning, then the Council of Industrial Design as education officer. He also taught at the Maidstone School of Art and the Royal College of Art.

Ironside collaborated with his brother Robin on design commissions, including stage designs for Sadlers Wells, decorations for Pall Mall during the Coronation, postage stamps and coins. The pair designed the obverse side of Britain's decimal coins as well as new coins for Tanzania, Brunei, Qatar, Dubai and Singapore.

He was appointed OBE in 1971.

Kerr, Sir (John) Graham
1869–1957
Zoologist

John Graham Kerr was educated at the Royal High School, Edinburgh, and the University of Edinburgh. Kerr interrupted his studies to join an expedition to study ornithology in Argentina (1889–91). His account of the expedition, *A Naturalist in the Gran Chaco*, was published in 1950.

Kerr returned to England in 1891 and attended Christ's College, Cambridge where he completed the Natural Sciences Tripos. He made a second expedition to Paraguay (1896–7). In 1897 he was appointed demonstrator in animal morphology at Cambridge and was a Fellow of Christ's (1898–1904). In 1902 he was appointed Regius Professor of Zoology at Glasgow, where he remained until 1935.

Kerr taught zoology to medical students throughout his professorship at Glasgow. He was also member of the court from 1913 to 1921 and served on the governing bodies of various other institutions. He was president of the Scottish Marine Biological Association (1942–9), a member of the advisory committee on fishery research from its foundation in 1919, and chairman (1942–9).

Kerr was closely involved in the development of general scientific activities in Scotland. He was elected Fellow of the Royal Society in 1909, served on the council (1920–2 and 1936–8), and was its vice-president (1937–8). He was also president of the Royal Physical Society of Edinburgh (1906–9), of the Royal Philosophical Society of Glasgow (1925–8), and was vice-president and Neill prize-winner of the Royal Society of Edinburgh in 1904. He served for many years on the council of the British Association and was president of the zoology section at the Oxford meeting in 1926.

He took a great interest in the application of correct biological principles in camouflage and on the outbreak of the First World War he wrote to the Admiralty advocating the use of disruptive patterning.

In 1935 Kerr was elected Member of Parliament for the Scottish Universities. He attended the House of Commons regularly and was for a time chairman of the parliamentary scientific committee. He was knighted in 1939.

La Dell, Edwin
1914–1970
Printmaker, painter and teacher

Edwin (T. E.) La Dell was educated at South Grove School. He won scholarships to train at Sheffield School of Art in 1930, then the Royal College of Art in 1935. Whilst at the RCA he joined a lithography class at the Central School of Arts and Crafts.

At the outbreak of the Second World War La Dell joined the Civil Defence Camouflage Establishment. He was sent on active service in 1943. During the war La Dell was involved with the Artists' International Association, a left-wing artists' group, which aimed to democratise fine art.

After the war La Dell returned to the Central School of Arts and Crafts to teach and, from 1948, taught lithography at the Royal College of Art, where he became head of the school of engraving in 1950.

He became a member of the Royal Society of British Artists in 1950 and was elected Associate of the Royal Academy in 1969.

Lewis, John Noel Claude
1912–1996
Typographer and graphic designer

John Lewis was educated at Charterhouse and trained as a graphic designer at Goldsmiths College. He worked as a freelance designer after leaving college.

Lewis joined the CD&TC in 1940 and was eventually put in charge of the newly opened South-Eastern Command Camouflage School in Tunbridge Wells. He was later posted to Canada to run the Canadian Army Camouflage School, before joining the 8th Army in Italy to inspect dummy tanks. On his return he rejoined the CD&TC as Talbot Kelly's assistant, becoming chief instructor on Kelly's retirement.

After the war Lewis practised as a graphic designer. He lectured at the Royal College of Art in the School of Graphic Design from 1951. He was invited to run the Lion and Unicorn Press at the RCA in 1956. Lewis left the RCA in 1964 to edit design books for Studio Vista, and to write several books of his own.

Autobiography: *Such Things Happen: The Life of a Typographer* (1994).

Maskelyne, Jasper

1902–1973
Magician

Jasper Maskelyne was born into a family of stage magicians. He and his stage partner George Cooke billed themselves as the Royal Illusionists and Anti-Spiritualists. He also was an inventor and patented the coin-operated toilet door. After Cooke's death Maskelyne formed a new partnership with another illusionist, David Devant.

During the Second World War, Maskelyne enlisted with the Royal Engineers and worked on deception schemes in North Africa.

After the war, Maskelyne resumed his stage career, but found that work as a stage magician had become scarce. He moved to Kenya where he ran a driving school.

Autobiography: (ghost-written) *Magic: Top Secret* (1949).

Biography (novelised): David Fisher, *The War Magician* (1983, reprinted in 2004).

Mason, Frank Henry Algernon

1876–1965
Marine illustrator, painter and printmaker

Frank Mason was educated as a cadet on HMS *Conway*. He spent a year at sea as an engineer, before starting an engineering apprenticeship, working for shipbuilders in Hartlepool and Scarborough. Despite his lack of formal training, Mason began a career as a painter in 1897/1898.

In 1914 he was commissioned to the Royal Naval Volunteer Reserve and in 1915 was posted to the Suez Canal. Back in the UK in 1916 he commanded a minesweeper operating off the northern coastline. Towards the end of the war, Mason was called up by the Admiralty to assist in naval camouflage as a dazzle painting officer, based first in Scotland then in Newcastle. He was demobilised in 1919.

Mason was commissioned after the war by the Imperial War Museum, to paint studies of Suez based on his wartime sketches. In the inter-war period he returned to work as a marine painter. He was commissioned to design posters for the railways, including the London and North Eastern Railway and, at the advent of the Second World War, a recruitment poster for the Royal Navy.

During the Second World War he joined the Directorate of Naval Camouflage, along with other marine painters Leslie Wilcox and Claude Muncaster. When the war ended he was 69 years old. He continued to tour and paint for the rest of his life.

Biography: Edward Yardley, *The Life and Career of Frank Henry Mason* (1996).

Medd, David Leslie

1917–
Architect

David Medd trained at the Architectural Association.

Medd served as an army camouflage officer at the Eastern Command Camouflage School in Norwich. From 1943 he was posted to the CD&TC, where he created prototypes for decoy and dummy devices.

After the war Medd worked in the architecture department of Hertfordshire County Council (1946–9), and then for the newly formed Architects and Building Branch of the Ministry of Education.

In 1949 he married the architect Mary Crowley. Both held strong views on child-centered learning and together they were responsible for designing educational buildings. They were influential in shaping the philosophy of the Architects and Building Branch at the Ministry.

He was appointed joint OBE with his wife Mary in 1964.

Medley, (Charles) Robert Owen

1905–1994
Stage designer, painter and teacher

Robert Medley was educated at Greshams School, Norfolk. He trained at Byam Shaw School of Art, the Royal Academy Schools, the Slade School of Fine Art, and took night classes under Bernard Meninsky at the Central School of Arts and Crafts.

He moved to Paris in 1926, studying at various art academies. On his return to London in 1930 Medley founded the Group Theatre with his partner Rupert Doone. At the Group Theatre Medley designed sets for T. S. Eliot's *Sweeney Agonistes* and W. H. Auden and Christopher Isherwood's *Dog Beneath the Skin*, *Dance of Death*, *Ascent of F6* and *On the Frontier*.

During the Second World War, Medley joined the CD&TC and in 1941 was posted to the Middle East, where he remained until the end of the war.

Medley taught painting at Chelsea School of Art (1932–9 and 1945–9). He established and ran the department of theatre design at the Slade School of Art (1949–1958), and was head of fine art at Camberwell School of Arts and Crafts (1958–65). From 1968 Medley was chairman of the faculty of painting at the British School in Rome.

He was elected Royal Academician in 1986, and appointed CBE in 1982.

Messel, Oliver Hilary Sambourne

1904–1978
Artist and stage designer

Oliver Messel was educated at Eton College, but left early to train at the Slade School of Fine Art. After leaving the Slade he was apprenticed to the portrait artist, John Wells.

In 1925 Messel showed a collection of masks at the 'Character Masks' exhibition at the Claridge Galleries. This led to his first commission, designing masks for ballet at the London Coliseum. He spent much of the 1930s designing for film, theatre and ballet.

When the Second World War began, Messel joined the CD&TC and after training was posted to the Eastern Command Camouflage School, Norwich.

After the war he continued to design costumes and scenery for theatre, opera and film until he moved to Barbados in 1966. He reinvented himself as an architect and interior designer on Barbados and Mustique.

Messel was made a Fellow of University College London in 1956, and was appointed CBE in 1958.

Morton, John

1919–
Architect and designer

John Morton initially trained as a teacher before enrolling at the School of Cabinet Making in Birmingham. He began working for the City of Birmingham Education Authority in September 1939, but was conscripted later that year as a carpenter in the Royal Army Ordnance Corps.

During the Second World War Morton served as a camouflage officer in the Middle East, returning to the UK in early 1944 when he transferred to RAF for pilot training.

After the war Morton entered the Architectural Association School and qualified in 1949. He was recruited by David Pye, and worked closely with R. D. Russell and R. Y. Goodden on the Lion and Unicorn Pavilion for the Festival of Britain (1951).

Moss, Colin William

1914–2005
Painter, critic and teacher

Colin Moss trained at Plymouth School of Art (1930–4) and the Royal College of Art (1934–8).

In 1938 he was commissioned to design murals for the British pavilion at the World's Fair in New York.

Shortly before the outbreak of the Second World War Moss joined the Air Ministry's programme to camouflage industrial buildings. When the war began he transferred to the Directorate of Camouflage. Moss was drafted into the Life Guards in 1941 and served in the Middle East as a captain, continuing in the Army Education Corps in Palestine after the war.

Moss was senior lecturer at Ipswich School of Art (1947–79). He studied under Kokoschka at the School of Vision in Salzburg in 1961.

Moss was founder member of the New Ipswich Art Group and the Six in Suffolk Group, and chairman of the Ipswich Art Club. From 1980 he was resident art critic for the *East Anglian Daily Times*.

Biography: Chloe Bennett, *Colin Moss: Life Observed* (1996).

Muncaster, Claude Grahame [formerly Oliver Grahame Hall]

1903–1974
Painter

Oliver Grahame Hall was educated at Queen Elizabeth's School, Kent. He had no formal art training but learned under the direction of his father. He adopted the name Claude Muncaster in 1922, formally changing it by deed poll in 1945.

During the Second World War Muncaster was attached to the Training and Staff Duties Division of the Admiralty's Directorate of Scientific Research, where he was put in charge of a team of design staff seconded from the Royal Naval Volunteers' Reserve and advised the Admiralty on the camouflage of ships at sea.

After the war Muncaster continued to paint. His commissions included a series of watercolours for Queen Elizabeth in 1946–7.

Muncaster was elected a member of the Royal Society of Painters in Water Colours in 1936, and president (1951–60). He was made a member of the Royal Society of British Artists in 1946, and of the Royal Institute of Oil Painters in 1948.

Biography: Martin Muncaster, *The Wind in the Oak* (1978).

Murray Dixon, William

1918–
Artist and designer

William Murray Dixon was educated at Oulton High School, Liverpool before winning scholarships to Liverpool City School of Art and the Royal College of Art, the latter under the professorship of Basil Ward. He was awarded First Prize for Architecture and Interior Design by the Federation of British Industry in 1939.

When the Second World War was declared Dixon was called up to the Rifle Brigade. He worked in army intelligence until he was promoted to sergeant and posted to the Western Desert. In 1943 Dixon transferred to the Royal Engineers and became art advisor to the chief education officer of troops at the Army Education Centre in Baghdad.

After the war Dixon returned to the RCA, graduating in design in 1947. He took up a position as a designer in De La Rue's plastics division, where he is credited with the development of Formica for use in furniture and exhibition design.

Dixon established a Formica design department and rose to the position of International Design Manager for Formica International and Formica UK. He developed the product for multiple uses, including 1.5 million square feet of patterned laminate specified for Cunard Line's flagship, the *Queen Elizabeth 2*.

Penrose, Sir Roland Algernon

1900–1984
Artist and writer

Roland Penrose was educated at Quaker Downs and Leighton Park schools. He went to Queens College, Cambridge to read history, switching to architecture in 1920. After graduating he went to Paris to study painting at the schools of Othon Friez and André Lhote.

On his return to England Penrose organised the International Surrealist Exhibition at the New Burlington Galleries (1936). In 1938 he established the London Gallery in Cork Street with E. L. T. Mesens, until the war forced its closure in 1940.

At the outbreak of the Second World War Penrose served as an ARP warden. He was later given a teaching post at the Home Guard Camouflage Unit, Osterley Park and became senior lecturer at the Eastern Command Camouflage School. He was a member of the Industrial Camouflage Research Unit.

After the war Penrose co-founded the Institute of Contemporary Arts – a space for new and experimental art of all kinds – with Mesens and Herbert Read (1947).

Penrose founded the Elephant Trust in 1975 together with Lee Miller, which was designed to support experimental artists and writers who might not gain funding elsewhere.

He was appointed CBE in 1961 and knighted in 1966.

Publications: *The Home Guard Manual of Camouflage* (1941).

Biography: Anthony Penrose, *Roland Penrose: The Friendly Surrealist* (2001); National Galleries of Scotland, *The Surrealist and The Photographer* (2001); Elizabeth Cowling, *Visiting Picasso: The Notebooks and Letters of Roland Penrose* (2006).

Proud, Ralph Priestman (Peter)

1913–1989

Art director

Peter Proud was educated at Hillhead High School.

He left school at 15 and found work at the British International Pictures studios at Elstree, initially as a sound recordist and then in the art department. He soon became assistant art director, working under Alfred Hitchcock on films such as *Murder!* (1930) and *Rich and Strange* (1932).

In 1932 Proud joined the art department at Gaumont-British where he co-art directed *The Man Who Knew Too Much* (Hitchcock, 1934), before moving to Warner Bros where he became head of the art department. He became known for an ability to make sets look as if they had been designed on lavish production budgets.

During the Second World War he joined the CD&TC, and was posted to the Middle East as second in command in charge of training and development.

After the war Proud joined Independent Producers. In the 1950s he moved into independent television production, where he used a new system for set design based on interchangeable modules.

In the 1980s, with over 70 productions to his name, Proud turned to education, teaching at the London International Film School until his death.

Pye, David William

1914–1993

Woodcarver, writer and teacher

David Pye was educated at Winchester College before training at the Architectural Association.

When the Second World War began he was attached to the Admiralty Training and Staff Duties Division, Directorate of Scientific Research to design camouflage.

After the war Pye taught at the Architectural Association for a short time before being invited in 1948 to teach furniture design at the Royal College of Art. He became a professor in 1963 and remained there until 1974.

Pye taught, designed and made furniture for much of the rest of his life.

He was appointed OBE in 1985.

Publications: *Ships* in the Penguin series *Things We See* (1950), *The Nature of Design* (1964), *The Nature and Art of Workmanship* (1968) and *The Nature and Aesthetics of Design* (1968).

Biography: *David Pye: Wood Carver and Turner*, Crafts Council in Association with the Crafts Study Centre (1986).

Ravilious, Eric

1903–1942

Painter, wood-engraver and designer

Eric Ravilious was trained at Eastbourne School of Art, then at the Royal College of Art.

His early work included a mural for the Morley College refectory, London, and wood engravings for the Golden Cockerel, Curwen and Nonesuch presses. In the 1930s he began to use lithography. Ravilious was also a designer for Wedgwood. His designs include the celebration mug for the coronation of George VI; the Alphabet mug (1937); Afternoon Tea (1937), Travel (1938) and Garden Implements (1939) china sets, and the Boat Race Day design (1938). He also designed glass for Stuart Crystal (1934), furniture for Dunbar Hay (1936) and advertising for London Transport.

Ravilious also worked in watercolour, painting rural English landscapes and interiors.

At the outbreak of the Second World War he became an official war artist. He died observing a sea rescue mission in 1942.

Robb, Brian
1913–1979
Illustrator, painter and teacher

Brian Robb was educated at Malvern College, Chelsea School of Art (1930–4) and the Slade School of Fine Art (1935–6).

During the Second World War, he was involved in camouflage and deception with the 8th Army in Italy and in the Middle East. His wartime cartoons were published in the magazines *Crusader* and *Parade*.

Robb taught at Chelsea School of Art (1936–62 before and after the war), and was head of the illustration department, Royal College of Art (1963–78).

Robb also designed publicity for Shell, Guinness, and *Punch*.

Publication: *My Middle East Campaigns*, 1944.

Russell, Richard Drew (Dick)
1903–1981
Architect and furniture designer

Richard Russell was educated at Dean Close School in Cheltenham. He left at the age of 16 to work as a furniture designer with his brother Gordon. He then trained at the Architectural Association (1924–8).

He became a director of Russell Workshops in 1928 (from 1929 called Gordon Russell Ltd). He designed clocks for Garrard, and became consultant designer to Murphy Radio. He returned to Gordon Russell Ltd in 1939.

During the Second World War, Russell was recruited into the Training and Staff Duties Department and worked in the Admiralty Naval Camouflage Unit.

He was appointed professor of wood, metal and plastics at the Royal College of Art in 1948, where he remained until 1964. He contributed furniture and interior design schemes to the 'Britain Can Make It' exhibition in 1943 and the Festival of Britain in 1951.

Russell remodelled the Greek Sculpture Galleries at the British Museum and, with Robert Goodden, its Western Sculpture, Print Room and Oriental Art Galleries (1969–71).

He was made Royal Designer for Industry in 1944.

Scott, Sir Peter Markham
1909–1989
Painter, naturalist and broadcaster

Peter Scott was educated at West Downs and Oundle schools, then studied at Trinity College Cambridge. Scott attended the Akademie der Bildenden Künste, Munich for a term, before spending two years at the Royal Academy Schools.

He produced his first book *Morning Flight* in 1935.

When the Second World War began, Scott volunteered for the Royal Naval Volunteer Reserve. He spent two years on destroyers, then served in the coastal forces in steam gunboats.

After the war Scott set up the Severn Wildfowl Trust, a research and conservation organisation (later known as the Wildfowl and Wetlands Trust).

Scott helped to establish the BBC Natural History Unit in Bristol and hosted the programme *Look* for 17 years. He also took part in the radio programme *Nature Parliament*.

Scott was involved in the International Union for the Conservation of Nature and Natural Resources from the early 1950s, and was its chairman (1962–81). In 1961 he co-founded the World Wildlife Fund, becoming its chairman and designing the famous panda logo.

He was rector of Aberdeen University (1960–3) and chancellor of Birmingham University (1974–83). He became both Chairman and Fellow of the Royal Society in 1987.

He was appointed MBE in 1942, CBE in 1953 and knighted in 1973.

Autobiography: *The Eye of the Wind* (1961), *Travel Diaries of a Naturalist* (1983–7).

Seago, Edward Brian (Ted)

1910–1974
Painter and writer

Edward Seago had little formal education due to a severe heart complaint. He studied painting with Bertram Priestman from the age of 13.

Seago toured England, Ireland and France with a travelling circus. He wrote and illustrated his circus travels in *Circus Company* (1933), *Sons of Sawdust*, (1934) and *Caravan* (1937).

During the Second World War he joined the CD&TC and was posted to Southern Command as a camouflage officer. He was invalided out of the Army in 1944 but returned to paint the final stages of the Italian campaign. His wartime writings include *Peace in War* (1943), *High Endeavour* (1944) and *With the Allied Armies in Italy* (1945).

He was elected to the Royal Society of British Artists in 1946 and the Royal Society of British Painters in Water Colours in 1959.

Biography: Jean Goodman, *Edward Seago: the other side of the canvas* (1978), Horace Shipp *Edward Seago: painter in the English tradition* (1952), Ron Ranson, *Edward Seago: a review of the years 1953–1964* (1992).

Spence, Sir Basil

1907–1976
Architect, exhibition and interior designer

Basil Spence was educated at George Watson's College, Edinburgh and trained, first in sculpture then architecture, at Edinburgh College of Art. He spent his year out at the office of Sir Edwin Lutyens and attended evening classes at the Bartlett School of Architecture (1929–30).

Spence returned to Edinburgh to complete his architectural training, joining the Edinburgh practice Rowand Anderson, Balfour Paul in 1935.

When the Second World War began Spence joined the CD&TC, rising to the position of major in the Camouflage Unit.

After the war the practice of Basil Spence & Partners was formed. The new practice designed exhibition pavilions for 'Enterprise Scotland' (1946), 'Scottish Industries' (1947) and 'Britain Can Make It' (1949), as well as the Sea and Ships Pavilion, Skylark Restaurant, Nelson Pier and the heavy industries stand for the Festival of Britain (1951). He also won the Festival of Britain award for public housing.

Spence opened a London office in 1951. Basil Spence & Partners continued to exist, but in addition an Edinburgh practice (Spence, Glover, and Ferguson) was formed in 1958 and a London practice (Spence, Bonnington, and Collins) two years later. In 1964 Spence also opened a small select office (Sir Basil Spence OM RA), which worked out of his London home.

Spence was president of the RIBA (1958–60) and Royal Designer for Industry in 1960.

He was knighted in 1960 and awarded the OM in 1962.

Stiebel, Victor

1907–1976
Couturier

Victor Stiebel arrived from South Africa in 1924 to study at Cambridge. Whilst at university he designed for theatre wardrobes. He then trained for three years in dress design under Reville & Rossiter before opening his own fashion house in 1932.

During the Second World War Stiebel joined the CD&TC, and was later posted to the Eastern Command Camouflage School. In 1942 Stiebel designed utility clothes for the Board of Trade made with a prescribed yardage of material, as well as uniforms for the WRENS and WRAF.

After the war he returned to design, working for Jacqmar and becoming chairman of the Incorporated Society of London Fashion Designers. He reopened his own house in 1958, initially with great success, but was forced to close on health grounds after only five years.

Stradling, Sir Reginald Edward

1891–1952
Civil engineer

Reginald Stradling was educated at Bristol Grammar School. In 1909 he won a Surveyors' Institution scholarship to study civil engineering at the

University of Bristol. He completed his practical training with consulting engineers API Cotterell, then worked for firms in Bolton and Birmingham.

At the outbreak of the First World War Stradling volunteered with the Royal Engineers, was commissioned and sent to France in 1915. He was invalided home in 1917 and awarded the Military Cross.

After the war Stradling taught civil engineering at Birmingham University, and was appointed head of civil engineering and building at Bradford Technical College in 1922. He resigned two years later to become director of building research at the Department of Scientific and Industrial Research. In 1937 his department collaborated with the Home Office Committee on Air Raid Precautions and problems associated with bomb damage.

In 1939 Stradling became chief adviser to the Ministry of Home Security, working on camouflage, smoke screening, air raid shelters, and the effects of Allied bombs on German targets.

After the war Stradling became chief scientific adviser to the Ministry of Works (1944–9) and adviser on civil defence to the Home Office (1945–8). He researched new building methods for post-war construction. In 1947 he became a member of the Advisory Council on Scientific Policy.

He was awarded the American Medal for Merit in 1947 for his work on the effects of explosions, received the James Alfred Ewing medal and was elected a Fellow of the Royal Society in 1943. He was member of the council of the Institution of Civil Engineers from 1939 until his death. In 1949 he became part-time dean of the Military College of Science in Shrivenham where he remained until his death.

He was appointed CBE in 1934 and knighted in 1945.

Sykes, Steven Barry
1914–1999
Artist, designer and teacher

Steven Sykes was educated at the Oratory School, and trained at the Royal College of Art.

Sykes won a travelling scholarship from the RCA, visiting France and Italy in the late 1930s. He then worked with Herbert Hendrie at his stained glass studio in Edinburgh.

When the Second World War broke out Sykes trained on No. 2 Camouflage Course with the CD&TC. He was posted as camouflage officer to North Africa and Egypt, then to France. He was promoted to major in 1941.

From 1946 Sykes taught at Chelsea School of Art, initially teaching perspective classes, before becoming head of design. He remained in this post until 1979.

Commissions included the Gethsemane Chapel in Basil Spence's Coventry Cathedral and ceramics for the Festival of Britain.

Biography: *Deceivers Ever: Memoirs of a Camouflage Officer* (1990).

Thayer, Abbott Handerson
1849–1921
Painter and naturalist

Abbott Thayer studied at the Brooklyn Art School, the National Academy of Design and, from 1875, the École des Beaux-Arts, Paris.

He returned to New York in 1879 and established himself as a portrait painter, early on making black and white illustrations of literary figures for *Scribner's Magazine*.

Thayer developed a theory of natural camouflage in animals, which appears in *Concealing Coloration in the Animal Kingdom* (1909), written by his son Gerald. A number of Thayer's theories were put to military use during the Second World War.

Biography: Nelson C. White, *Abbott H. Thayer: Painter and Naturalist* (1951), Alexander Nemerov, *Vanishing Americans: Abbott Thayer, Theodore Roosevelt and the Attraction of Camouflage* (1997).

Trevelyan, Julian Otto
1910–1988
Painter, printmaker, writer and teacher

Julian Trevelyan was educated at Bedales School (1923–8) and Trinity College, Cambridge from 1928. He left in 1930 and moved to Paris to study at Atelier 17.

Trevelyan joined the English Surrealist Group (resigning in 1938), the

Artists' International Association in 1937, and participated in Mass-Observation (1937–8). During the war, he helped to form the Industrial Camouflage Research Unit before joining the CD&TC, where he served as a camouflage officer mainly in North Africa. He was invalided out of the Army in 1943.

Trevelyan taught etching at Chelsea School of Art (1950–60), and engraving at the Royal College of Art (1955–63).

He was elected Honorary Senior Royal Academician in 1986.

Autobiography: *Indigo Days* (1957).

Wadsworth, Edward Alexander

1889–1949
Painter and printmaker

Edward Wadsworth was educated at Fettes College in Edinburgh. On leaving school he lived in Munich (1906–7) to learn German and machine draughtsmanship. He also studied printing, woodcutting, painting, and non-technical drawing. In 1908 Wadsworth went to Bradford School of Art, where he won a scholarship to the Slade School of Fine Art (1909–12).

In 1913 he joined Roger Fry's decorative arts co-operative, the Omega Workshop. He resigned and, together with Wyndham Lewis, formed the rival group, the Rebel Art Club. The group met to discuss revolutionary art ideas, leading to the birth of Vorticism.

During the First World War, Wadsworth served with the Royal Naval Volunteer Reserve as an intelligence officer on the Greek island of Mudros, before being employed to develop dazzle camouflage.

He moved away from Vorticism in the 1920s when the sea became a more important subject in his work. In the 1930s he was associated with abstract artist group, Unit 1, established by Paul Nash, Wells Coates, Henry Moore and Ben Nicholson, and the constructivist magazine, *Circle*, published by Ben Nicholson and Naum Gabo. In 1938 Wadsworth was commissioned to design a mural for the saloon of the Cunard Lines' *Queen Mary*.

Biography: Barbara Wadsworth, *Edward Wadsworth: A Painter's Life* (1989).

Wilkinson, Norman

1878–1971
Marine artist, printmaker and poster designer

Norman Wilkinson was educated at Berkhamsted and St Paul's Catholic Choir School, Portsmouth. He studied figure painting in Paris, then trained with the river and coastal painter Louis Grier. He later attended the Southsea School of Art where he returned to teach. He began his career as an illustrator on the *Illustrated London News* in 1898.

In 1915 he served in the Navy and produced on-the-spot drawings for his book *The Dardenelles*. Wilkinson made an important contribution to the development of camouflage techniques. He invented dazzle painting in the First World War, and during the Second World War was elected Honorary Air Commodore, in which capacity he advised the Air Ministry on camouflage. In 1944 he presented 54 paintings from a series entitled 'The War at Sea' to the nation.

Wilkinson also designed posters for the London and North Western Railway, and the London Midland and Scottish Railway – organising a series of posters by Royal Academicians for the latter in 1924.

He was appointed CBE in 1948.

Autobiography: *A Brush with Life* (1969).

Publications: *The Dardanelles* (1915), *Colour Sketches from Gallipoli* (1915), *Ships in Pictures* (1945).

Yunge-Bateman, James

1893–1959
Painter, wood engraver and teacher

James Yunge-Bateman trained as a sculptor in Leeds (1910–14). He won a scholarship to the Royal College of Art, but was called up to serve in the First World War. Injuries sustained during the war forced him to give up sculpture for painting, which he studied at the Slade School of Fine Art (1919–21). Yunge-Bateman taught at Cheltenham School of Art (1922–8), then Hammersmith School of Art.

During the Second World War Yunge-Bateman became director of naval camouflage at Leamington Spa.

182

Glossary

21 Army Group

GHQ 21 Army Group was a closely integrated force set up by General Montgomery to mastermind and implement plans for Operation Overlord and D-Day. It had no dedicated training centre. Its camouflage units were formed in 1943 from various experienced branches of the Royal Engineers, with the Camouflage Pool and Units kept under close central control.

Admiralty

The government department responsible for administering the Royal Navy.

Air Ministry

The government department charged with managing the affairs of the Royal Air Force.

The Allies

During the Second World War a group of countries officially opposed the Axis Powers (Nazi Germany, Fascist Italy and the Empire of Japan). The Allies comprised the United States of America, the Soviet Union, and the United Kingdom (including the Indian Empire and Crown Colonies). China joined following the attack on Pearl Harbour on 7 December 1941.

The Architectural Association School of Architecture

The AA was founded in the mid-nineteenth century as an independent organisation addressing the education and professional training of young architects. As a school as well as an association formed by a worldwide membership of individuals committed to architectural learning and knowledge of all kinds, the AA has evolved in parallel with the realities that have shaped the architectural profession during the last 50 years.

Artists' International Association

An association of left-wing British artists founded in London in 1932, the AIA aimed for 'the unity of artists against Fascism and war and the suppression of culture'. Originally it was called Artists International, but the word 'Association' was added to its name on its reconstitution in 1935. The Association continued until 1971, but abandoned its political objectives in 1953, thereafter existing as an exhibiting society. A series of large group exhibitions based on political and social themes was held, beginning in 1935 with the exhibition 'Artists Against Fascism and War'. The AIA supported the left-wing Republicans during the Spanish Civil War (1936–9) through exhibitions and other fund-raising activities. It tried to promote wider access to art through travelling exhibitions, public murals and affordable lithographs such as the Everyman Prints series produced in 1940.

The Arts Council

The Arts Council of Great Britain (known since 1994 as Arts Council England, the Scottish Arts Council and the Arts Council of Wales) was created in 1946 as the successor to the Committee for the Encouragement of Music and the Arts (CEMA), which was formally set up by Royal Charter in 1940. The Arts Council was accorded Royal Charter status in 1946.

CEMA evolved as a scheme to improve national morale and provide employment for artists during wartime. The Arts Council continued CEMA's efforts to make the arts accessible to the general public but, whereas CEMA provided the arts directly, through promoting theatre and concert tours, the Arts Council gave more of these responsibilities to arts organisations. The Arts Council's objectives were to assist and encourage the improvement of professional standards and the distribution of arts throughout the country, also to delegate local responsibility for promoting the arts, and to provide buildings for arts-related activities.

Barrage balloon
A balloon anchored singly or in a series over a military objective to support nets designed to hinder the passage of enemy aircraft.

The blackout
The British Government imposed periods of compulsory darkness on the populace when air raids took place during the Second World War.

The Blitz
The devastating German air raids on Britain carried out during 1940 and 1941.

'Britain Can Make It'
Three months after the end of the Second World War the Council of Industrial Design staged a major exhibition of consumer goods entitled 'Britain Can Make It'. The exhibition was intended both to lift the morale of the British public with a vision of a new future, and to boost export trade in British products in order to pay for post-war reconstruction. Contributors were briefed to exhibit manufactured, rather than hand-made goods.

'Britain Can Make It' opened at the Victoria & Albert Museum in September 1946. It flaunted the best of British industrial production of consumer goods for the home and garden, as well as jewellery, fashion, personal accessories, dress fabrics, toys, interior decoration, packaging, home entertainment, sports equipment, and furniture. Nearly 1.5 million people visited the exhibition, and orders worth between £25 and £50 million were generated.

British 8th Army
An Allied formation of the Second World War, which fought in campaigns in North Africa and Italy. The 8th Army was formed from the Western Desert Force in September 1941. The formation included divisions from the Australian and Indian Armies, also South Africans, New Zealanders and a brigade of Free French.

Central Institute of Art and Design
An influential body set up in 1939 with the object of supporting Britain's artists, coordinating the interests of approximately 40 organisations connected with art and design. It promoted schemes for the employment of artists in times of war.

Coastal Command
An organisation within the Royal Air Force which defended the United Kingdom from naval threat, and countered German submarines by air.

Command (n)
A team of officers exercising control over a particular operation.

The Committee of Imperial Defence
The Government body responsible for researching defence strategies between 1904 and the start of the Second World War. During the First World War, its duties were largely taken over by the War Council, and it did not resume full operations until 1922. Typically, a temporary sub-committee would be set up to investigate and report in detail on a specific issue.

The Corps of Royal Engineers
The Corps of Royal Engineers originated in the sixteenth century in the early days of the Ordnance Office, when military engineers were employed for duty in Royal arsenals and fortifications. In 1717 a permanent officer corps of engineers was established, the prefix 'royal' being added in 1787. In 1856 the Corps of Royal Sappers and Miners was absorbed into the Royal Engineers, bringing together engineering corps of both officers and soldiers. The Royal Engineers became responsible for all technical developments and functions within the army.

Council of Industrial Design (COID)
Established with the help of a generous grant by the Board of Trade in December 1944, the COID was responsible for both the 'Britain Can Make It' exhibition (1946) and the Festival of Britain (1951). In 1956 the Design Centre was opened, thus providing a permanent venue in which to display quality goods, and in 1972 the COID itself was reformulated as the Design Council.

Dazzle camouflage
A style of naval camouflage developed in the First World War which aimed at deception rather than obscurity.

D-Day
6 June 1944, the historic day on which the Allied forces invaded northern France.

Disruptive camouflage
A style of camouflage using broken-up surfaces in order to confuse the eye.

The Enigma codes
Encrypted German military messages were successfully broken by cryptologists at Bletchley Park.

The Establishment
A slang term for the traditional Conservative governing body. The word has come to imply a strong resistance to change.

The Festival of Britain
The Festival of Britain was held in 1951 as a celebration of national recovery, and to mark the centenary of the Great Exhibition of 1851. London's South Bank was the main exhibition site, and included the Festival Hall, the Dome of Discovery, the Skylon and Shot Tower. It was a triumphant display of the design, technology and industry of post-war Britain.

British designers and architects created a series of themed Pavilions, such as Power and Production, Transport and Communications, Homes and Gardens, and Sport.

The Home Guard
In May 1940 a Government message was broadcast asking for volunteers for the LDV (Local Defence Volunteers) to defend the Home Front. In August 1940 the name was changed to the Home Guard. Within a month of the broadcast, 750,000 men had volunteered, and by the end of June 1940 the total number of volunteers stood at over 1 million. The Home Guard was instructed to 'stand down' at the end of 1944. The Home Guard was never tested by an invasion and it operated under inevitable constraints.

Key Points Intelligence
Key Points Intelligence originated in the Intelligence Section of the Air Ministry's Camouflage Branch, where it compiled lists of 'target' factories requiring defence. It was transferred to the Ministry of Home Security in September 1939, and became the Key Points Intelligence Branch. It gathered and distributed information about air raid damage to vital factories, public utilities, railways and buildings of national importance. It was designated a Directorate in September 1942.

Merchant ships
Ships carrying commercial or crucial wartime supplies.

The Ministry of Information
The central government department responsible for publicity and propaganda in the Second World War.

The Munich Crisis
On 29 September 1938 a four-power conference comprising senior statesmen from Britain, France, Germany and Italy took place in Munich. At this meeting it was agreed that Nazi Germany would take charge of Sudetenland, a border area of Czechoslovakia mainly populated by ethnic Germans. In return for this, Hitler promised not to make any further territorial demands on Europe.

The Munich Agreement was signed, transferring Sudetenland to Germany. Although the agreement was popular in Britain, as it appeared to have prevented a war with Germany, major politicians including Churchill and Eden attacked the agreement, indicating that Britain had behaved dishonourably and, furthermore, lost the support of the illustrious Czech army. In March 1939 Hitler broke the agreement and seized the rest of Czechoslovakia, thereby beginning the major hostilities which were the outbreak of the Second World War.

Official War Artist Scheme
During the Second World War the War Artists Advisory Committee, which was chaired by Sir Kenneth Clark, began a scheme to record the war effort through official artists' commissions. Over 300 artists were involved, including John Piper, Graham Sutherland, Henry Moore, Eric Ravilious and Stanley Spencer on the home front, and Anthony Gross, Edward Bawden and Edward Ardizzone overseas. The pictures collected were exhibited both in London and in shows which toured nationally and internationally. In 1946 one-third of the collection was allocated to the Imperial War Museum.

Office of Works
The former title given to the Ministry of Works, which was formed in 1943 to organise the requisition of property for wartime use.

Osterley Park
During the Second World War the grounds of this Middlesex stately home were commandeered by Tom Wintringham and used as a training school for the Home Guard.

Pillbox
A static concrete guard post positioned on the coast, or close to road junctions, rivers or railways. Most pillboxes comprised a small room of about 10 feet square and six feet high, with thick, rough concrete walls with a look-out, and narrow slits for machine guns.

Radar
A system for detecting the presence, direction, distance and speed of aircraft, ships, and other craft, by emitting pulses of radio waves which are reflected off the object back to the source.

The Royal College of Art
The Royal College of Art was founded in 1837 as the Government School of Design, to educate young designers in order that they could work in the manufacturing industries. In 1853, as the National Art Training School, it moved to South Kensington as part of Prince Albert's cultural development of the area. In 1896 the school was renamed the Royal College of Art, in recognition of its broadening opportunities. In 1948 the College reopened after the Second World War with courses in many new disciplines, and it was awarded its Royal Charter in 1967. It continues to uphold its unique status as the world's only wholly post-graduate university of art and design.

RDI
The Royal Designer for Industry award was established by the RSA in 1936 to encourage high standards of industrial design and to boost the status of designers. It is awarded to designers from a broad range of fields, who have achieved 'sustained excellence in aesthetic and efficient design for industry'.

Royal Naval Volunteer Reserve
The RNVR was established in 1903 and consisted of those who had civilian jobs but were part-time sailors in their spare time. Members were drafted as ratings or officers in both world wars. Records of their work are held at The National Archives, and the Fleet Air Arm museum.

Royal Ordnance Factories
The collective name given to the Government's Second World War munitions factories built to maximise the capacity of existing factories situated in, or close to, London and therefore vulnerable to attack. New ordnance factories were built in areas considered to be 'relatively safe'.

186

The Royal Society for the Encouragement of Arts, Manufactures and Commerce (RSA)
A multi-disciplinary institution founded in 1754 by William Shipley, a painter and social activist, 'to embolden enterprise, enlarge science, refine arts, improve our manufactures and extend our commerce'. Today the RSA's work is focused on five manifesto challenges, 'enterprise, environment, education, communities and citizenship', which reflect the original mission in twenty-first century terms.

The Slade School of Fine Art
The Slade School was founded in 1871 as the result of a bequest from Felix Slade, who envisaged a school where fine art could be studied within a liberal arts university. The school is located in University College London. It continues to uphold its tradition of studio-based taught programmes of contemporary art and the practice, history and theories which inform it.

Society of Industrial Artists (SIAD)
A professional body for designers founded in 1930 and originally entitled the Society of Industrial Artists. From 1960 it became the Society of Industrial Artists and Designers and in 1986 it was renamed the Chartered Society of Designers.

U-Boat
An abbreviation of Unterseeboot, referring to the German naval submarines deployed during the First and Second World Wars.

Vorticism
A British avant-garde group formed in London in 1914 by the artist and writer Wyndham Lewis. Vorticist imagery was typically fragmented and hard-edged, and celebrated the dynamic modern world, the machine and the urban environment. Ezra Pound and T. S. Eliot contributed to the movement's journal *BLAST*, which embodied the movement's manifesto.

War Cabinet
A committee formed by a Government in time of war. When he became Prime Minister in May 1940, Winston Churchill formed a War Cabinet, which initially included Neville Chamberlain, Clement Attlee, Lord Halifax and Arthur Greenwood.

War Office
A former Government department responsible for the administration of the British Army between the seventeenth century and 1963, when its functions were transferred to the Ministry of Defence.

Western Desert Campaign
The primary early theatre of the North African Campaign of the Second World War. In Egypt the German Afrika Corps posed a new threat in the Western Desert under the aggressive leadership of General Rommel. Then, in June 1942, Rommel captured the important port of Tobruk. Two months later, General Montgomery took command of the British 8th Army and ended its series of defeats with the victory at El Alamein. Operation Torch, the landing of three British and American armies under the control of General Eisenhower, led to the surrender in May of the Axis forces in North Africa, the capture of Sicily, and the subsequent invasion of Italy.

Western Front
The Second World War theatre of fighting west of Germany, encompassing France, Belgium, the Netherlands, Luxembourg, and Denmark.

Zeppelin
A large German dirigible airship in service during the early twentieth century.

Select Bibliography

Allwood, Rosamund and Laurie, Kedrun, *R. D. Russell and Marian Pepler* (London, Inner London Education Authority, 1983).

Barkas, Geoffrey, *The Camouflage Story* (London, Cassell, 1952).

Behrens, Roy, *Art and Camouflage: Concealment and Deception in Nature, Art and War* (Iowa, University of Northern Iowa, 1981).

Behrens, Roy, *False Colors: Art, Design and Modern Camouflage* (Iowa, Bobolink Books, 2002).

Bennett, Chloe, *Colin Moss: Life Observed* (Ipswich, Malthouse Press, 1996).

Blechman, Hardy, *DPM: Disruptive Pattern Material: An Encyclopaedia of Camouflage* (London, DPM Publishing, 2004).

Brandon, Laura, et al., *Shared Experience: Art and War* (Canberra, Australian War Memorial, 2005).

Brentnall, Margaret and Hutton, Marigold, *John Hutton, Artist and Glass Engraver* (Philadelphia, The Art Alliance Press, 1986).

Buckton, Henry, *Artists and Authors at War* (Barnsley, Pen and Sword, 1999).

Chesney, C. H. R., *The Art of Camouflage* (London, Robert Hale, 1941).

Cott, H. B., *Adaptive Coloration in Animals* (London, Methuen, 1940).

Dobinson, Colin, *Fields of Deception: Britain's Bombing Decoys of World War II* (London, Methuen, 2002).

Fletcher, Alan, *The Art of Looking Sideways* (London, Phaidon Press, 2001).

Frayling, Christopher, *The Royal College of Art: One Hundred and Fifty Years of Art and Design* (London, Barrie & Jenkins, 1987).

Frayling, Christopher, *Art and Design: One Hundred Years at the Royal College of Art* (London, Richard Dennis, 1999).

Gardner, James, *The ARTful Designer: Ideas off the Drawing Board* (London, Centurion Press, 1993).

Hartcup, Guy, *Camouflage: A History of Concealment and Deception in War* (Newton Abbott, David & Charles, 1979).

Lewis, John, *Such Things Happen: The Life of a Typographer* (Stowmarket, Unicorn Press, 1994).

MacCarthy, Fiona, *British Design Since 1880: A Visual History* (London, Lund Humphries, 1982).

MacCarthy, Fiona and Nuttgens, Patrick, *Eye for Industry: Royal Designers for Industry 1936–1986* (London, Lund Humphries, 1986).

Manser, José, *Hugh Casson: A Biography* (London, Viking, 2000).

Maskelyne, Jasper, *Magic: Top Secret* (London, Stanley Paul, 1949).

Matthew, H. C. G. and Harrison, Brian, eds., *The Oxford Dictionary of National Biography* (Oxford, Oxford University Press, 2004).

Morris, Linda and Radford, Robert, *AIA: The Story of the Artists' International Association, 1933–1953* (Oxford, The Museum of Modern Art, 1983).

Muncaster, Martin, *The Wind in the Oak: The Life, Work and Philosophy of the Landscape Artist Claude Muncaster* (London, Robin Garton, 1978).

Newark, Tim, Newark, Quentin and Borsarello, J.F., *Brassey's Book of Camouflage* (London, Brassey's, 1986).

Penrose, Antony, *Roland Penrose: The Friendly Surrealist* (London, Prestel, 2001).

Penrose, Roland, *The Home Guard Manual of Camouflage* (London, Routledge, 1941).

Penrose, Roland, *Scrap Book: 1900–1981* (London, Thames & Hudson, 1981).

Reit, Seymour, *Masquerade: The Amazing Camouflage Deceptions of World War 11* (London, Robert Hale, 1979).

Robb, Brian, *My Middle East Campaigns* (London, Collins, 1944).

Scott, Peter, *The Eye of the Wind* (London, Hodder & Stoughton, 1961).

Spalding, Frances, ed., *Ravilious in Public* (Fitzroy, Black Dog Books, 2002).

Sykes, Stephen, *Deceivers Ever: Memoirs of a Camouflage Officer, 1939–1945* (Speldhurst, Spellmount, 1990).

Trevelyan, Julian, *Indigo Days* (London, MacGibbon & Kee, 1987).

Ullmann, Anne, ed., *Ravilious at War* (Upper Denby, The Fleece Press, 2002).

Who Was Who, 1929–1940, 2nd edn, Vol 3 (London, A&C Black, 1967).

Williams, D, *Liners in Battledress* (St Catherines, Vanwell Publishing, 1989).

Williams, D, *Naval Camouflage 1914–1945* (London, Chatham, 2001).

189

Index